The Palgrave Macmillan Animal Ethics Series

Series editors: **Andrew Linzey** and **Priscilla Cohn**

In recent years, there has been a growing interest in the ethics of our treatment of animals. Philosophers have led the way and a range of other scholars have followed, from historians to social scientists. From being a marginal issue, animals have become an emerging issue in ethics and in multidisciplinary enquiry. This series explores the challenges that animal ethics poses, both conceptually and practically, to traditional understandings of human–animal relations.

Specifically, the series will

- provide a range of key introductory and advanced texts that map out ethical positions on animals;
- publish pioneering work written by new, as well as accomplished, scholars; and
- produce texts from a variety of disciplines that are multidisciplinary in character or have multidisciplinary relevance.

Titles include:

Aysha Akhtar
ANIMALS AND PUBLIC HEALTH
Why Treating Animals Better Is Critical to Human Welfare

Alasdair Cochrane
AN INTRODUCTION TO ANIMALS AND POLITICAL THEORY

Andrew Knight
THE COSTS AND BENEFITS OF ANIMAL EXPERIMENTS

Claire Molloy
POPULAR MEDIA AND ANIMAL ETHICS

Siobhan O'Sullivan
ANIMALS, EQUALITY AND DEMOCRACY

Thomas Ryan
SOCIAL WORK AND ANIMALS: A MORAL INTRODUCTION

Kay Peggs
ANIMALS AND SOCIOLOGY

Joan Schaffner
AN INTRODUCTION TO ANIMALS AND THE LAW

Forthcoming titles:

Mark Bernstein
HUMAN–ANIMAL RELATIONS: THE OBLIGATION TO CARE

Eleonora Gullone
ANIMAL ABUSE AND HUMAN AGGRESSION

Alastair Harden
ANIMALS IN THE CLASSICAL WORLD: ETHICAL PERCEPTIONS

Lisa Johnson
POWER, KNOWLEDGE, ANIMALS

The Palgrave Macmillan Animal Ethics Series

**Series Standing Order ISBN 978–0–230–57686–5 Hardback
978–0–230–57687–2 Paperback**
(*outside North America only*)

You can receive future titles in this series as they are published by placing a standing order. Please contact your bookseller or, in case of difficulty, write to us at the address below with your name and address, the title of the series and the ISBN quoted above.

Customer Services Department, Macmillan Distribution Ltd, Houndmills, Basingstoke, Hampshire RG21 6XS, England

Animals and Sociology

Kay Peggs
University of Portsmouth, UK

 Montante Family Library
D'Youville College

© Kay Peggs 2012

All rights reserved. No reproduction, copy or transmission of this publication may be made without written permission.

No portion of this publication may be reproduced, copied or transmitted save with written permission or in accordance with the provisions of the Copyright, Designs and Patents Act 1988, or under the terms of any licence permitting limited copying issued by the Copyright Licensing Agency, Saffron House, 6–10 Kirby Street, London EC1N 8TS.

Any person who does any unauthorized act in relation to this publication may be liable to criminal prosecution and civil claims for damages.

The author has asserted her right to be identified as the author of this work in accordance with the Copyright, Designs and Patents Act 1988.

First published 2012 by
PALGRAVE MACMILLAN

Palgrave Macmillan in the UK is an imprint of Macmillan Publishers Limited, registered in England, company number 785998, of Houndmills, Basingstoke, Hampshire RG21 6XS.

Palgrave Macmillan in the US is a division of St Martin's Press LLC,
175 Fifth Avenue, New York, NY 10010.

Palgrave Macmillan is the global academic imprint of the above companies and has companies and representatives throughout the world.

Palgrave® and Macmillan® are registered trademarks in the United States, the United Kingdom, Europe and other countries.

ISBN 978–0–230–29257–4 hardback
ISBN 978–0–230–29258–1 paperback

This book is printed on paper suitable for recycling and made from fully managed and sustained forest sources. Logging, pulping and manufacturing processes are expected to conform to the environmental regulations of the country of origin.

A catalogue record for this book is available from the British Library.

A catalog record for this book is available from the Library of Congress.

10 9 8 7 6 5 4 3 2 1
21 20 19 18 17 16 15 14 13 12

Printed and bound in Great Britain by
CPI Antony Rowe, Chippenham and Eastbourne

Contents

Series Preface vi

Acknowledgements viii

1. Sociology and Animals — 1
2. Animals and Biology as Destiny — 16
3. Animals, Social Inequalities and Oppression — 33
4. Animals, Crime and Abuse — 47
5. Town and Country: Animals, Space and Place — 65
6. Consumption of the Animal — 90
7. Animals, Leisure and Culture — 107
8. Animal Experiments and Animal Rights — 127
9. Conclusions: Sociology for Other Animals — 145

References 153

Author Index 166

Subject Index 170

Series Preface

This is a new book series for a new field of enquiry: Animal Ethics.

In recent years, there has been a growing interest in the ethics of our treatment of animals. Philosophers have led the way and a range of other scholars have followed, from historians to social scientists. From being a marginal issue, animals have become an emerging issue in ethics and in multidisciplinary enquiry.

In addition, a rethink of the status of animals has been fuelled by a range of scientific investigations which have revealed the complexity of animal sentience, cognition and awareness. The ethical implications of this new knowledge are yet to be properly evaluated, but it is becoming clear that the old view that animals are mere things, tools, machines or commodities cannot be sustained ethically.

But it is not only philosophy and science that are putting animals on the agenda. Increasingly, in Europe and the United States, animals are becoming a political issue as political parties vie for the 'green' and 'animal' vote. In turn, political scientists are beginning to look again at the history of political thought in relation to animals, and historians are beginning to revisit the political history of animal protection.

As animals grow as an issue of importance, so there have been more collaborative academic ventures leading to conference volumes, special journal issues and indeed new academic animal journals as well. Moreover, we have witnessed the growth of academic courses, as well as university posts, in Animal Ethics, Animal Welfare, Animal Rights, Animal Law, Animals and Philosophy, Human–Animal Studies, Critical Animal Studies, Animals and Society, Animals in Literature, Animals and Religion – tangible signs that a new academic discipline is emerging.

'Animal Ethics' is the new term for the academic exploration of the moral status of the nonhuman, an exploration that explicitly involves a focus on what we owe animals morally, and which also helps us to understand the influences – social, legal, cultural, religious and political – that legitimate animal abuse. This series explores the challenges that Animal Ethics poses, both conceptually and practically, to traditional understandings of human–animal relations.

The series is needed for three reasons: (i) to provide the texts that will service the new university courses on animals; (ii) to support the increasing number of students studying and academics researching in animal-related fields and (iii) because there is currently no book series that is a focus for multidisciplinary research in the field.

Specifically, the series will

- provide a range of key introductory and advanced texts that map out ethical positions on animals;
- publish pioneering work written by new, as well as accomplished, scholars; and
- produce texts from a variety of disciplines that are multidisciplinary in character or have multidisciplinary relevance.

The new *Palgrave Macmillan Animal Ethics Series* is the result of a unique partnership between Palgrave Macmillan and the Ferrater Mora Oxford Centre for Animal Ethics, UK. The series is an integral part of the mission of the Centre to put animals on the intellectual agenda by facilitating academic research and publication, and is also a natural complement to one of the Centre's other major projects, the *Journal of Animal Ethics*. The Centre is an independent 'think tank' for the advancement of progressive thought about animals, and is the first of its kind in the world. It aims to demonstrate rigorous intellectual enquiry and the highest standards of scholarship. It strives to be a world-class centre of academic excellence in its field.

We invite academics to visit the Centre's website at www.oxfordanimalethics.com and to contact us with new book proposals for the series.

Acknowledgements

Thanks to all those who have contributed to this book in one way or another. Colleagues and students from the School of Social, Historical and Literary Studies at the University of Portsmouth, UK, have provided a rich environment for thought and discussion. My colleagues in sociology deserve a special mention. In particular, I would like to thank the students in my Animals in Society classes with whom I discussed each of the chapters – I wish I had the space to name all of you. I am also grateful to the Faculty of Humanities and Social Sciences and the Centre for European and International Studies Research at the University of Portsmouth, UK, for granting workload relief. My special thanks go to Andrew Linzey who suggested that I write this book instead of the one I originally proposed. I am also grateful to Clifton P. Flynn for his comments on the entire draft and to the anonymous referees who pointed me in directions that I had not thought of. The editorial team at Palgrave Macmillan has been excellent; my special thanks to Priyanka Gibbons, Melanie Blair and Cherline Daniel. Remaining errors and mistakes are, of course, entirely my own. I cannot end without saying thank you to Tracey Jones and Ian Parry who, in addition to being wonderful neighbours, were able to tell me where donkeys live in Hampshire, UK. Lastly, but most importantly, I have discussed all of my ideas with Graham Attenborough; his thoughts about nonhuman animals in past and present societies are a constant inspiration.

1
Sociology and Animals

Introduction

In his book *The Compleat Observer*, Jack Sanger (1996, p. 5) recounts the story of a man who crosses the border between two countries each day, with a donkey pulling a cart full of straw. The border guard notices that the man looks wealthier each time he crosses and so concludes that the man must be smuggling. The next time the man tries to cross the border the guard orders the straw to be searched for smuggled goods, but nothing is found. On subsequent crossings he orders the straw to be cut into pieces, to be boiled and finally to be burned, but still the guard finds nothing. Meanwhile, the man gets wealthier and wealthier. Eventually the guard gives up. Some years later the guard bumps into the man, who has now retired to a beautiful house in the country. Astounded, the guard asks him to come clean and tell him how he became so rich. Since the man is no longer smuggling, he is happy to own up; he used to smuggle donkeys. Sociology has acted rather like the guard in this story in largely concentrating on what it expects to find rather than opening itself up to the possibilities of what might be out there, and in doing so has often overlooked other animals.

In examining other animals in society this book provides an exploration of major issues in sociology, and in doing so challenges assumptions about the very nature of sociology. Such assumptions are evident in the definitions of sociology that are often found in introductory sociology textbooks. Any sociology textbook would do as most definitions seem to be essentially the same. For example, in their very good introductory textbook, John Macionis and Ken Plummer (2005) define sociology as 'the systematic study of human society'. What should strike us about this definition is that it seems to dispute that there can be a focus on other animals by sociology; such a focus challenges what is seen as the core of sociological study, which is human society. This is not to say that sociologists have left unchallenged the established view that sociology's remit is the study of only humans. For example, in the 1970s Clifton D. Bryant stressed that 'Our social enterprise is not composed of humans alone' (1979, p. 417) as, he observed, other animals are everywhere in our lives. Because human societies are infused with other animals, he argued that sociology could gain a great deal from investigating this observable reality. For example, often humans eat the flesh of other animals and wear their skins and hair as clothing. Other animals live with humans in their homes and work for humans in a myriad of ways. Human entertainment often centres on the performances of other animals and human speech often invokes other animal metaphors. Thus, for Bryant (1979), other animals are central to the study of society and for this reason he called on sociologists to recognize the important roles that other animals take in human societies.

It is curious that sociology has conventionally sought to limit the field of sociological inquiry to humans alone because sociology is proud of its otherwise wide sphere of interest. To be sure, hardly any (if any) feature of human society seems to be beyond the remit of sociology and indeed, John O'Neill observes, 'Sociology owes its fortune to the fact that nothing occupies man like himself' (1972, p. 3). Leaving aside the considerable discussion we might have about gender in response to O'Neill's assertion, the important roles that other animals play in human societies have been largely ignored and human relations with other animals have been seen as marginal to sociology's main focus (Kruse, 2002). In consequence, Olin Myers argues, this has led to 'possible blind spots in sociological theory' (2003, p. 46). We can find evidence of such omissions in George Herbert Mead's (1934) assertion that other animals are outside of the realm of sociological enquiry because of their purported lack of perception, imagination and language. But, sociology did not always overlook other animals; for example, the classical sociologist Max Weber suggested that human relations with other animals might be an appropriate subject matter for sociology (Myers, 2003, p. 49). However, Mead's subsequent position came to dominate and it was not until the 1970s that Bryant (1979) called on sociologists to recognize the roles of other animals in human societies. Although other animals are still seen as marginal to sociology, this is not to say that there has not been some progress. This chapter introduces the debate about the place of other animals in sociology. Of course, it is not possible to cover everything: sociology is an enormous field of study. Thus, I have been unavoidably selective. However, in this chapter, and in those that follow, I hope the important contribution that sociology can make to such study will become evident. My discussion throughout centres on the significant contribution that

- sociology can make to our understanding of human relations with other animals,
- sociology can make to understanding relations among other animals, and
- the study of other animals in society can make to sociology.

In setting out the debate this introductory chapter provides a backdrop to those that follow by exploring sociology as it is usually conceived. As we will see, the 'doing' of sociology is grounded in using the sociological imagination (Mills, 1970), which encourages us to question the way things seem to be. The sociological imagination is central to this chapter, and to this book, as it is this imagination that encourages us to see issues such as human relations with other animals as political issues and thus, paradoxically, to question and criticize conventional understandings of what sociology is. This is not a bad thing as, Robert Bierstedt declares, criticism of one's own discipline is a commendable exercise (1960, p. 3). Nevertheless, we will temper criticisms by examining how the acceptance of other animals in sociological analysis enhances rather than challenges the richness and diversity of the discipline.

What is sociology?

We have already seen how Macionis and Plummer define sociology; let us take another definition. In their equally good introductory text, Robin Cohen and Paul Kennedy suggest that 'Sociology involves the systematic study of patterns of human interaction' (2000, p. 3). The similarity between this definition and that offered by Macionis and Plummer is obvious; the focus is on the human. The resemblance is not surprising as such definitions are standard in a range of sociology textbooks. I aim to use this standard definition as a means by which we can review debates about what the 'appropriate' subject matter of sociology is and what is the 'proper' way of doing sociology. Standard definitions emphasize the systematic or 'disciplined' approach, as Peter Berger (1963) describes it, to the study of society. This emphasis brings to mind the scientific approach advocated by Auguste Comte, who first coined the term 'Sociology'. Comte described sociology as the scientific study of society as it really is (1975 (orig 1851–54)). The introductory textbooks cited so far make clear that the crucial thing is that Comte sought to distance sociology from previous studies of society that were grounded in theological approaches (which saw society as an expression of God's will) and metaphysical approaches (which saw society as a natural phenomenon) as, he argued, such nonsociological approaches concentrated on how society *should be*. He wanted to promote a scientific study of society that would reveal what he claimed to be the unchanging laws that govern the way that society *actually is* (Comte, 1975 (orig 1851–54)). Comte saw the development from theological through metaphysical to scientific approaches to knowledge as a three-stage law of social and intellectual progress. He likened his sociological approach to the approach taken in natural science, that is as a 'social physics' composed of invariable laws. Consequently, Comte was positivistic in his approach as he believed that natural scientific approaches are the only appropriate methods for gaining what he saw as *true* knowledge about society. However, since Comte was writing, sociology has become a more multifaceted enterprise and this is clear, not least, in the range of methodological approaches adopted by sociologists. Alternative approaches are often grounded in Weber's (1978 (orig 1925)) contention that human behaviour cannot be explained by rigid laws and, for that reason, the study of societies must address the subjective meanings of those who live within them. Importantly for this book, as we will see below, one of the upshots of sociology's concern with subjective meanings has been the establishment of a deep division between humans and other animals in sociology. However, before we turn to this let me consider first the role that the sociological imagination takes in how we 'do' sociology.

The sociological imagination

Rather than being purely an academic study, sociologists are engaged in a way of seeing that, to draw on Berger, amounts to seeing familiar things as strange, because familiar things are 'not what they seem' (1963, p. 34). For example, in *The Sociological Imagination* Charles Wright Mills (1970) challenges us to look at occurrences (we might use the example of eating 'meat') as the consequence of social forces (such as conventions about acceptable and unacceptable food products) rather than as the result of individual choice. If Mills had referred to the role of other animals in society, he might have argued that we should look at eating their bodies as food in terms of agribusiness rather than in terms of individual food preferences; I will come to this later in Chapter 6. What is important here is that Mills argued that we should not be deceived into seeing behaviour in terms of merely what people decide to do; rather we should see behaviour as guided by society. So, sociology encourages us to challenge that which is taken-for-granted, and in this way for sociologists the sociological imagination becomes part of their daily lives rather than merely an aspect of a dry academic pursuit. Consequently, once we develop our sociological imagination we cannot put it aside. In enabling us to challenge that which is taken-for-granted, the sociological imagination could enable us to become, in Mills' (1970) terms, more active participants in society. This is because the sociological imagination encourages us to grow more aware of social ills, such as suffering and inequalities, and encourages us to want to change the world to improve it (though we might well have different ideas about what constitutes an 'ill' and what signifies an 'improvement'). Thus, Steve Fuller suggests, sociology has sought to achieve the ambition of creating a 'heaven on earth', which could be defined as 'a world in which humans exercise dominion over nature without exercising dominion over each other' (2006, p. 1). Nevertheless, if we return to Comte, this sociological enterprise of achieving 'heaven on earth' seems to recall the type of approach that he rejected; he wanted to abandon study of society that focused on 'what should be' in favour of study that centres on 'what is'. The important division between sociology's role in 'what should be' and 'what is' will be returned to most notably in Chapter 9 when we consider what Beth Humphries (1997) calls the 'emancipatory potential of sociology'. For now I want to turn my attention to a crucial division in sociology, which is between humans and other animals. If we look again at Fuller's definition of the ambition of sociology, it brings to mind this division because it is the assumed superiority of humans over all other aspects of nature that makes humans the singular focal point of sociology.

Humans in sociology

Thinking back to the definitions of sociology presented so far it is clear that the focus is on *humans* in what are seen as *human* societies. As I have said, such definitions are standard. In the glossary to the sixth edition to his textbook *Sociology*, Anthony Giddens (2009) opens his definition of sociology with the statement that it is 'The study of human groups and societies...'. The emphasis on humans found in this and the previous definitions cited conforms to what Janet Alger and Steven Alger call 'the hard line that sociology has always drawn between humans and other species' (2003, p. 69). As we will see below, other animals are rarely mentioned in sociology textbooks and this serves to reflect and to shape the view that other animals are outside of the remit of sociology. Indeed, it was only in 2002 that the American Sociological Association (ASA) granted full section status to an 'Animals in Society' section and, as Corwin R. Kruse notes, this caused some disquiet as some sociologists claimed that sociology should focus on humans alone (2002, p. 375). Despite the disquiet, sociologists who believed that animals are appropriate for the study of sociology gained ground and in June 2006 The British Sociological Association (BSA) inaugurated its own 'Animals/Human Study Group'. Nevertheless, both the study of human relations with other animals and the study of relations between other animals are viewed by many sociologists as marginal to sociology (for discussion see Arluke, 2002). Why have some sociologists taken this view? To answer this question I turn to Mead's notion of the gulf that exists between humans and other animals.

Sociological perspectives that are rooted in action, especially the perspective of symbolic interactionism, can be used to explore and explain the deep division between humans and other animals that has come to dominate sociology. The work of Max Weber, Herbert Blumer and George Herbert Mead is especially central in this respect. Mead's interpretation of humans as profoundly distinct from other animals seems to be accepted as the origin of the fundamental division between humans and other animals in sociology (e.g. see Alger and Alger, 2003). Although today there are a number of symbolic interactionist studies that focus on human relations with other animals (e.g. Arluke and Hafferty, 1996; Sanders, 1999; Twining and Arluke, 2000), Mead saw no room for other animals in his approach to sociology. Why is this so? In order to answer this question it is useful to turn to the origins of Mead's interactionist approach.

Action and symbolic interaction

The study of action is now fundamental to sociology, and the classical sociologist Max Weber is often seen as being the originator of this approach. In his reference to 'action', Weber centred on behaviour to which individuals attach a subjective meaning (1978 (orig 1925), p. 4). He felt that the study of meaning is fundamental to the study of society and so, Weber argued, sociology should concern 'itself with the interpretive understanding of social action' (1978 (orig 1925), p. 4). Weber's sociology, with its focus on meaning, is quite different from the sociological approach advocated by Comte. Although, like Comte, Weber (1978 (orig

1925)) defined sociology as 'a science', their approaches to this scientific study differed considerably. Weber argued that sociology should examine how human meanings and actions shape society, Comte (1975 (orig 1851–54)) maintained that sociology should concern itself with identifying invariable laws that govern the way that society is. I will come to differing ideas about objectivity in sociology in Chapter 9; here my focus is on the study of meaning and action as being central to sociology. So far I have only referred to individual action and the meanings that people attach to their behaviours. Weber argued that 'action is "social" because the way in which individuals behave and the subjective meanings they attach to their behaviours take into account the views and behaviours of others' (1978 (orig 1925), p. 4). For Weber, social reality is composed of individuals who take actions that are determined by, for example, goals, values, feelings and habits, and it is this that is the subject matter of sociology. This approach led to the development of symbolic interactionism.

Herbert Blumer coined 'symbolic interactionism' to describe an action approach that was a development of earlier theories. Symbolic interactionism draws attention to the role of interpretation in the attachment of meaning to actions (Blumer, 1969). Language is central to Blumer's (1969) analysis as, he argued, the symbolic medium of language enables individuals to negotiate meaning. What he means by this is that by speaking with each other (i.e. through *symbolic interaction)* humans come to recognize meanings. Sociologists remain divided about whether sociology should centre on explanations of social reality based on meaningful action or on structural forces, and we will return to this later in the book. What is important in this chapter is that Blumer's focus on language is dependent on Mead's work on language which has been seen as heralding the deep division between humans and other animals in sociology.

George Herbert Mead and language

George Herbert Mead considered humans to be unique in their widespread and imaginative use of language and crucially, he argued, the human use of language is central to the development of the human sense of self (1934, p. 74). This is crucial to the absolute distinction that Mead made between humans and other animals and the related disassociation of sociology from the study of other animals. Mead argued that humans are unlike other animals because, through interaction with others, humans develop a self which is different from the purely biological being that defines other animals (1934, p. 74). Human interaction depends on shared meanings (as we cannot communicate with each other if we do not understand things in similar ways), and Mead felt that this sharing of meanings is a distinctive feature of human societies. Although he concedes that other animals can embark on meaningful actions (such as the pursuit of other animals) that are designed to attain goals (such as food), he argued that the behaviour of other animals 'lacks the premeditation and shared meaning that characterize human behaviour' (Irvine, 2003, p. 46). He saw the actions of other animals as instinctual actions that are based on what he called a 'conversation of gestures'. He argued that the 'conversation of gestures does not carry with it a symbol which has a universal significance to all the different individuals' (1934, p. 55). In order to explain this idea, Mead uses the example of a fight between two dogs. In such a fight, each dog's reaction to what the other does is not one of conscious communication in anticipation of the other dog's behaviour, rather, Mead suggests, each dog is participating in a conversation of gestures in which each dog readjusts his or her position and attitude in response to each other's gestures (1934, pp. 42–3). Thus the dogs are involved in unconscious communication. Humans also participate in conversations of gestures. Mead cites the example of a mother's response to her baby crying, which, he suggests, is unconscious communication associated with protection. However, Mead argues, the human use of language (which he calls the use of 'significant symbols') supersedes this unconscious communication. Mead states:

> The importance of what we term 'communication' lies in the fact that it provides a form of behaviour in which the organism or the individual may become an object to himself. It is that sort of communication which we have been discussing – not communication in the sense of the cluck of the hen to the chickens, or the bark of a wolf to the pack, or the lowing of a cow, but communication in the sense of significant symbols, communication which is directed not only to others but also to the individual himself.
>
> (1934, p. 137)

I will come back to the idea of the individual as an object to herself or himself below, but let me remain with language for a moment. What Mead is interested in is how words come to provoke the same ideas in two or more people. For Mead, it is only language that produces a common definition in both the supplier of the words (the writer or speaker) and in the recipient of the words; accordingly, it is language that enables humans, and only humans, to anticipate the consequences of their actions, evaluate alternatives and organize their actions with others (1934, p. 47). So if we hear a word like 'fire', to use Leslie Irvine's (2003) example, we generate mental images of the situation, our position in it and what to do. Although Mead (1934) conceded that birds, to use one example, have a repertoire of sounds found in their songs unlike humans, he suggests, birds do not hold an image in their brains that they associate with each note. This is important to us here because, for Mead, it is only via conversation through significant symbols that participation in the social is possible (1934, p. 47).

This brings me to Mead's notion of the individual as an object to herself or himself. The ability to imagine is central to Mead's claim about the uniqueness of the human, because it is through this ability to imagine ourselves in a situation that we are able to see our 'self' as if it were an external object. For Mead this attribute 'demonstrates our [human] evolutionary advancement on animals' (1934, note 1). Accordingly, of other animals he says, 'We say an animal does not think... he does not put himself in the place of the other person' (1934, p. 73). Humans, he suggests, do put themselves in the place of another person through the perspective of the 'generalized other'. By this Mead is referring to the ways in which a person's idea of the expectations that others have about behaviour enables them to imagine the behaviour that is expected of them. For example, a sociologist who thinks it would be morally right to do research from an 'animal advocacy' perspective might be loath to do so because they feel that this might be deemed to be in conflict with sociological expectations of objectivity (we will come to the substance of this in Chapter 9). In thinking like this the sociologist would be taking on the perspective of the generalized other which enables them to think about how to act, when to act and the consequences of their actions. Taking into account the attitudes of other people is essential to acting 'intelligently' or 'rationally', and for Mead this is the province of humans alone (1934, p. 137). This is because intelligent or rational action does not rely on instinct; on the contrary such action relies on the capacity to make choices, and choices can only be made where there is a sense of the past that provides evidence of the relative consequences of different actions. Although we might think we see these features in other animals, for Mead we are merely making anthropomorphic projections as, he argues, other animals have 'no mind, no thought, and hence there is no meaning [in their behaviour] in the significant or self-conscious sense' (Mead, 1964, p. 136). So, for Mead, other animals act on instinct alone and live only in the present.

In summary, it is the purportedly unique social capacities of humans that Mead saw as the subject matter of sociology and, although this has not gone unchallenged, this position has largely remained. In consequence, 'sociologists...are supposed to study people, not other creatures' (Kruse, 2002, p. 375). The focus on humans is obvious in the content of sociology textbooks. Alger and Alger (2003) spent a good deal of time reading through 30 introductory sociology textbooks that were published in the United States between 1998 and 2002 and on the market in December 2001. They were interested in whether other animals were mentioned and, if they were, how they were constructed. Of the 30 textbooks, 27 referred to other animals, though there was considerable variation in the coverage (Alger and Alger, 2003, p. 71). They found that, where mention was made, the discussion reiterated rather than repudiated Mead's categorical differences between humans and other animals. They conclude:

> ...in our view, then, the authors of most introductory texts, like Mead, misconstruct animals as inferior, in order to construct humans as superior. Part of this comes from inadequate scholarship but much of it comes from various patterns of disparagement and denial.
> (Alger and Alger, 2003, pp. 74–5)

As an example of such disparagement, Alger and Alger point to what they call the 'distancing concepts' that are used in the textbooks to distinguish humans from other animals (2003, p. 81). They give the example of 'instinct', which was constructed in over a third of the textbooks as being applicable to the behaviour of other animals but not to that of humans; humans are constructed as uniquely without instincts (I discuss the role of instinct in Chapter 2). So, when other animals are mentioned in sociology textbooks they are often negatively constructed. I will explore the implication of this in more detail in subsequent chapters, but before we leave this general introduction I want to draw attention to the suggestions made by Bryant about the ways in which sociologists could include other animals in their studies.

Incorporating the societal role of other animals: Clifton D. Bryant

In 1979 Bryant observed that, in one way or another, human societies are infused with other animals and thus, he argued, sociology could achieve a great deal from investigating this observable reality.

He pointed to considerable scope for this as, he suggested, 'Sociological research on animal-related human behaviour might well yield additional valuable insights concerning the interactional process, social motivation, the influence of value systems on perception, socialization and personality development, human violence and its sublimation, and the social dynamics of anthropomorphism' (1979, pp. 404–5). With this in mind he suggested that sociology could examine a range of issues associated with, for instance, the other animal as a 'social problem', the other animal as a sentient creature, the linkages between humans and other animals at work, the other animal as surrogate human and zoological crime (which would include sociological work on other animals as personal property and as public property) (Bryant, 1979, pp. 405–6). He uses the term 'the zoological connection' to encapsulate this influence but, he notes:

> In spite of the evident prominence of zoological influences in our culture and the subsequent import for our social lives, the sociological literature is largely silent on animal-related human behaviour. This is an unfortunate oversight which handicaps our acquisition of a comprehensive understanding of our social enterprise.
> (Bryant, 1979, p. 404)

In consequence, he calls on sociologists to 'turn our sociological attention to this neglected area of social causation' (Bryant, 1979, p. 417). This book aims to do just that. Although my take on other animals and sociology differs from that taken by Bryant (I come to this in Chapter 9), I support his fundamental message: other animals are much neglected in sociology. This book aims to explore the ways in which sociology embraces, and can further develop, significant and worthwhile analyses of this important field.

Outline of the book

In the following chapters I aim to do a range of things. So far, I have only referred to action and symbolic interactionist perspectives in sociology; in the chapters that follow, these and other perspectives (covering both classical and contemporary sociological theories) will be explored. In returning time and again to the 'sociological imagination' I aim to provide insights into how sociological analysis can be useful in understanding a range of issues that are pertinent to the sociology of all animals (both human and nonhuman) in society.

I begin with biology. Sociology has done much to challenge biological deterministic ideas about differences among humans, and in Chapter 2 I explore the sociological challenge to biology and critique naturalistic approaches that see biological composition of beings as a justification for inequalities. Naturalistic assumptions have a longer history than sociology, thus I look back to the work of the seventeenth-century philosopher Rene Descartes and come forward to present-day thinking on the cognition of other animals. Developments in genetics have been used to reinforce previous theories about differences between species, but paradoxically such developments have also shown species' similarities in genetic make-up; so developments in genetics will also be central to the discussion. The chapter closes by offering an argument that is at variance with that put forward by Mead by examining sociological work that contends that other animals do indeed have selfhood.

In sociology, the examination of inequalities and oppression centres on the power relations, inequalities and disadvantages that are fundamental to most social systems. Despite the prevalence of inequalities associated with being 'animal', sociological analysis has largely centred on power relations among humans. Drawing on David Nibert's (2002) theoretical framework for explaining the origins of the oppression of humans and other animals, in Chapter 3 I explore how other animals have been (and can be further) drawn into sociological analyses of social inequalities and oppression. In addition to examining social divisions associated with, for example, class, 'race' and gender, the chapter draws on sociological thinking about difference, similarity and 'otherness' as central aspects of stratification and oppression.

In Chapter 4 I centre on crime and the abuse of other animals. Although the perpetrators of the most heinous of crimes are often referred to as 'animals', it is humans who are the executors of such acts, and humans and other animals who are the victims. Using the work of, for example, Piers Beirne (2002), this chapter explores the abuse of other animals through the lens of the sociology of crime, and in doing so examines the ways in which perpetrators of crime are 'animalized', the ways in which other animals are at risk from humans (e.g. through neglect and cruelty) and the ways in which humans use other animals for hazardous purposes (e.g. as status symbols and weapons in gang violence). Drawing on the work of Clifton P. Flynn, among others, the chapter also explores a developed area of sociology, that is the sociological examination of the association between the abuse of other animals and the abuse of humans.

Using the sociology of space and place, in Chapter 5 I explore spatial relationships between humans and other animals and examine how these are related to power relations between humans and other animals. I ask: how are other animals included and excluded from different places? In order to answer this question I compare urban areas such as towns and cities with areas viewed as being more natural and consider how humans have distinguished between places for other animals and places not for other animals (or specific other animals). In order to do this I reflect on the labelling of other animals as 'wild' and 'domesticated' and consider a range of segregated places, such as zoos, farms, laboratories and slaughterhouses, and integrated places, such as people's homes. To put this into context I begin with a brief discussion about how sociology has treated space and place.

A major way in which humans experience other animals in their everyday lives is through the consumption of food. Many humans consume the flesh of other animals, eat products such as eggs and milk, and consume related products such as cheese and butter. Not only are pieces or derivatives of other animals in many of the foods that humans consume but they are also a major component of what humans wear. Using the example of 'meat' consumption, in Chapter 6 I explore the ways in which other animals have become consumer products. The sociology of consumption provides insights into the gross materialism (Ritzer 1993) that defines societies in the west, and this gross materialism is an essential part of the McDonaldization thesis. In considering George Ritzer's (1993) notion of the McDonaldization of society I draw on work that sees links between the Fordist approach advocated by McDonald's and the rise of vegetarianism.

People often associate leisure with the freedom to relax and recuperate. Eating 'meat'-based foods when eating out, going horse-riding and going 'fishing' are obvious ways in which humans use other animals for the purposes of leisure. Other animals are also used for human entertainment, for example living animals are used as spectacle in zoos and representations of other animals appear in books, paintings and films. Such representations tell us a good deal about who we think we are, as these representations often function to distinguish us from other animals. Using Steve Baker's notion of disnification, Chapter 7 explores why other animals are often represented in comical ways and whether, when we are looking at the 'real' rather than the 'representation', we are ever seeing the real thing.

Much of the discussion has centred on the ways in which humans exploit and abuse other animals. In Chapter 8 I explore grass-roots activism as it is associated with experiments on other animals. The broad movement that is against such experiments is often labelled in terms of 'rights'. In this chapter I examine different conceptualizations of 'animal rights' as it is discussed in the philosophical literature (which is nuanced in its approach) and as it is more commonly understood by the public and in sociology. This enables consideration of the differing positions taken by animal advocates who put their energies into a range of group activities. The chapter considers the differences between the rights approaches and draws on a counter-movement that advocates experiments on other animals in order to introduce my notion of 'primacy identity politics' (Peggs, 2009a) as one way of exploring counter-arguments to animal rights.

Bryant's (1979) argument for a zoological connection in sociology is confirmed in Chapters 1–8; societies are indeed broader than the human. Sociology is part of the re-visioning of humans, in our looking for and seeing other animals. Still, is this re-visioning merely about describing the facts of human relations with other animals, or does sociology have a role to play in countering the oppression of other animals? Discussion about this question will close the book as in Chapter 9 I consider the purpose of sociology as it might relate to other animals.

Like those who have written about other animals in society, during the writing of this book I battled with terminology. I had to come to some decisions. Here and elsewhere I use the term 'other animals' to refer to animals who are nonhuman. I had a difficult choice to make with this term. Nibert (2002) makes clear that the language we use often obscures oppression and the hierarchical ranking of creatures. I can see what Nibert means when he suggests that the term 'humans and other animals' emphasizes 'human commonality with other inhabitants of the planet' (2002, p. xv); however, I feel that 'other animals' is still problematic as 'human' remains the referent. Like with the term 'nonhuman' the other creatures so labelled are linguistically constructed as 'other than us'. Nevertheless, I do not think 'other animal' is any worse than 'nonhuman' and I ended up choosing 'other animal' because I think it flows better. Like Nibert, I feel bothered by the oversimplified categorization of the numerous and diverse groups who I have labelled as 'other animals' (2002, p. xv), yet, like him, I have not found a more considerate and courteous way to express this diversity. I have thought good and hard about terms like 'hunting' and 'meat'. Although I agree that such terms obscure rather than make plain what is really going on (i.e. 'hunting' is chasing and murdering another animal and eating 'meat' is consuming the flesh of a dead animal), uses of alternatives can also obscure meaning. What I have decided to do is flag up the problems and assumptions contained in such words when I first use them and then contain the words in quotation marks if I have to use them from then on. This might make the pages look messy at times, but this is the compromise I have made.

2
Animals and Biology as Destiny

Introduction

Sociology has been regarded traditionally as the study of humans in societies. Sociological theories that centre on action, which stress the role of meaning in social interaction, point us to the *social* rather than the *biological* foundations of sociological analysis. Because other animals have been, and continue to be, associated with the biological rather than with the social it has been argued (by Mead (1934) and others) that the appropriate focus in sociology is on humans alone. However, Bryant (1979) argues that sociology has a great deal to offer in the study of the 'zoological connection'. In this chapter I take this further by beginning to explore issues associated with other animals in society as they might be examined in sociology. Because sociological theories point us in the direction of the social rather than the biological, I begin by exploring sociological thinking on the role of biology. Although the history of the development of sociology and biology has been entwined (Fuller, 2006, p. 80), this entanglement has at times been less than harmonious, not least because sociology has sought to challenge biological deterministic notions about differences among humans. In this chapter I explore the challenge that sociology has made to biology and critique naturalistic approaches that view the biological composition of beings as a justification for inequalities.

Naturalistic assumptions and binary classifications

A good starting point for thinking about assumptions about natural divisions among species is to try to answer the question 'what are the broad types of living things?' We might begin by answering 'plant–animal'. This reply points to one of the ways in which we answer such questions;

answers often take the form of labels and we often label things in terms of binaries, namely in two-fold terms. Gender is a good example as we think of gender in terms of the binary division of female–male. Such binaries often can be subdivided into further binaries. So, if we go back to our reply to the first question, that is 'plant–animal' (we might call this a first level binary), we can take the element 'animal' and ask a further question, 'what is an animal?' (this we might think of a second level binary). In all probability humans were included in the label 'animal' in our first reply but in our subsequent reply we might distinguish 'human–animal'. Now things are starting to get messy because human is at the same time an animal (in the first level binary) and not an animal (in the second level binary). Let us say this is just a problem associated with convenience, since we do not want to keep talking about 'lower animals' and 'humans' (which is how the Chambers Dictionary (1994) distinguishes the two) when we communicate with each other. Nevertheless, the messiness remains. The inconsistency we have found in the human–animal binary is not the end of the problem. A major problem with binaries is that they do not seem to be taken-for-granted labels at all; rather they seem to be fixed labels that are tied to unalterable natural objects. Thus, even if we forget about the inconsistency that is caught up in the binary labels 'human'-'animal', we nevertheless tend to think of the objects so labelled as being naturally occurring. The ideas of Rene Descartes have been a powerful influence on the way in which we perceive the 'natural' difference between humans and other animals (Dupre, 2004, p. 218). Indeed during Descartes' time humans were not thought of as animals at all, on the contrary humans were thought of as thinking beings while animals were thought of as machines (Rowlands, 2002).

Biological distinctions: Human and nonhuman

An essential difference between humans and other animals is established in the perception that humans, unlike other animals, have transcended their biology; it is precisely this transcendence that is said to define the human as human. Descartes is considered to be enormously influential on this way of thinking, and his views are influential to this day.

Rene Descartes: Animals as machines

The French philosopher Rene Descartes is proclaimed to be one of the founders of modern science. He is credited with shaping the view that 'humans' and 'animals' are distinctly different. Thus Descartes would not have referred to *other* animals; rather for him there *are humans* and there *are animals*. This demarcation is established in his belief that 'the mind is distinct from and superior to matter' (Sutcliffe, 1968, p. 19). Descartes came to this conclusion through the use of the methodological approach of observation, an approach that rejects ideas that cannot be verified by direct observation (Kalof and Fitzgerald, 2007, p. 59). Descartes argued that, unlike human behaviour, the observed behaviour of all other animals can be explained without attributing minds and consciousness to them (Armstrong and Botzler, 1993, p. 281). He believed that the universe is composed of two kinds of substances: the incorporeal mind (i.e. the mental substance of the mind) and the corporeal soul (i.e. the material substance of the body). He explains:

> ... one is purely mechanical and corporeal and depends solely on the force of the spirits and the construction of our organs, and can be called the corporeal soul: the other is the incorporeal mind, the soul which I have defined as a thinking substance.
> (Descartes, 1993, p. 284)

He saw humans as being composed of both mind and body (both the incorporeal mind and the corporeal soul); while he likened other animals to automata or machines as he believed that they are mindless material substances (they are simply corporeal souls). Consequently, for Descartes '... all the things which dogs, horses and monkeys are taught to perform are only expressions of their fear, their hope, their joy; and consequently they can be performed without any thought' (2007, p. 60). In distinct contrast, Descartes felt that the human use of language and reason clearly displays the presence of thought and this demarcates humans from other animals (and the rest of the natural world) (Armstrong and Botzler, 1993, p. 281).

Descartes' ideas clearly influenced Mead (Chapter 1), though it seems unlikely that Mead would have agreed with Descartes' opinion that other animals cannot feel pain (though some dispute that Descartes actually took this opinion (e.g. see discussion in Francione, 2008)). Conclusions of Descartes are grounded in his observation that, unlike humans, other animals do not speak. Because Descartes felt that speech demonstrates the presence of mind he concluded that other animals do not have minds. He insists that

> ... although all animals easily communicate to us, by voice or bodily movement, their natural impulses of anger, fear, hunger and so on, it has never yet been observed that any brute animal reached the

stage of using real speech, that is to say, of indicating by word or sign something pertaining to pure thought and not to natural impulse. Such speech is the only certain sign of thought hidden in a body. All men use it, however stupid and insane they may be, and though they may lack tongue and organs of voice: but no animals do. Consequently it can be taken as a real specific difference between men and dumb animals.

(Descartes, 1993, p. 285)

Accordingly, for Descartes, communication among other animals is disassociated from thought because it is purely automatic and unconscious. He bases this assumption on his distinction between three different grades of sensation: physical, conscious and self-conscious (Regan, 1983). For Descartes, humans have only the first (i.e. physical) sensation in common with other animals; the other two sensations (which are expressions of thinking) are the preserve of humans. Although humans and other animals physically respond to stimuli, for Descartes the response from other animals is a mechanical one because, unlike humans, other animals do not consciously feel pain and other sensations 'in the strict sense'. Consequently he viewed other animals as being no more than machines with parts assembled in intricate ways. As other animals are like machines Descartes argued that humans have little responsibility to them, unless the treatment of them affects other humans. For him, this 'absolves them [humans] from the suspicion of crime when they eat or kill animals' (Descartes, 2007, p. 62). This will be an important matter when, in later chapters, we come to think about issues associated with, for example, crime and violence (Chapter 4), the consumption of animal products such as 'meat' (Chapter 6) and experiments on other animals (Chapter 8). Although Descartes was not the first to assert that such natural differences exist between humans and other animals, our current thinking about the fundamental nature of these differences stems from Descartes' ideas (Dupre, 2002). Because a sentient being is one who is aware, conscious and capable of sensation (Francione, 2008), for Descartes other animals are not sentient beings. In consequence, in the seventeenth and eighteenth centuries, Cartesian scientists undertook experiments on living (other) animals who were fully awake throughout (Rowlands, 2002, p. 3). For sure, the Cartesian view that other animals do not feel pain conflicts with our present day thinking about sentience; however, the Cartesian perspective is clear in the current idea that *some* other animals (e.g. most invertebrates) do not feel pain. Thus, for other animals in the latter group, the proposed lack of sentience absolves humans of mistreatment of them. So, sentience has been used both as a basis for biological difference and as a basis for moral consideration. We will discuss the implications of this in subsequent chapters.

Why are understandings of biological differences so important to us in this book? In contrast to biology, which is defined as the study of all living organisms (from molecules, through cells to the whole organism), sociology is defined as the study of social life. This has meant that 'sociologists often react with hostility to explanations that evoke biology' (Freese et al., 2003, p. 233), and the exclusion of biology from the subject matter of sociology can be found in works by such eminent sociologists as Marx, Durkheim and Weber (Fuller, 2006, p. 81). The sociologist Guang Guo explains that the absence of explicit discussion of biological influences in sociology is established in at least two factors: '(1) our discipline's key theoretical emphasis on a group's social-structural position and (2) the unavailability of reliable measures of relevant biological influences' (2006, p. 145). But implicit in this is the overriding sociological assumption that humans are born the same, and that the differences among them are attributable to individual positions in the social hierarchy (Guo, 2006, p. 145). Using gender as an example, hierarchical differences between women and men are not biologically determined and naturally occurring, rather they are social creations based in unequal power relations. The separation between biology and sociology is important in this book, not least because if we agree with Descartes that other animals do not have minds and with Mead's assertion that animals are simply biological beings without society, then other animals cannot be the focus of sociological study. Two assumptions lie behind this proposition. Firstly, it is assumed that the division between humans and other animals is purely biological and thus this 'natural' division-making is outside of the remit of sociology. Secondly, because 'Culture consists of the phenomena studied by the social scientists' (Dupre, 2004, p. 897) and it is assumed that other animals are not really important to human culture and do not have their own culture, their worlds are not relevant for sociological analysis either. This is paradoxical position to take in sociology because, once we apply the sociological imagination, we can think about the ways in which

- the distinction between humans and other animals is also cultural
- other animals are embedded in human culture
- other animals have culture and have selves.

So the sociological imagination enables us to imagine and explore the study of other animals as an important aspect of the subject matter of sociology. We will see later in this chapter that the points above are not mutually exclusive; however, before we get there first I want to explore biological determinism, as this is central to sociology's rejection of the explicitly biological.

Biological comparisons: Biology as destiny

Biological essentialism is a significant issue in sociology because sociology has spent much time countering biological essentialist notions of the biological, physiological and (increasingly discussed) genetic roots of human social behaviour. Biological essentialists give little weight to sociological or cultural explanations of difference. For example, biological essentialists might argue that women are programmed to be more caring than men and men are programmed to be more aggressive than women, and because this particular social difference and/or behaviour is biological in essence it is unchangeable. This is a very crude account of biological essentialism as there is an array of biological essentialist views (Dupre, 2004), but there is not sufficient space to detail them here. However, what I can say here is that the broad brush of biological essentialism continues to shape popular conceptions of difference and similarity (though these conceptualizations are being challenged on a regular basis). As suggested above, these popular conceptions can be seen in naturalistic assumptions about the roots of gender inequalities and in terms of the focus in this book: differences between humans and other animals. Although sociologists have tended to react against biologically essentialist views associated with, for instance, gender inequalities, biologically essentialist thinking about the differences between humans and other animals have tended, for the large part, to go unchallenged (Chapter 1). Nevertheless, this is not to say there has been sociological silence on this matter, and counter-arguments to biological essentialism that concern gender, for example, are often very useful for considering assumptions about the human distinction from other animals. Sociobiological views of the world have tried to challenge sociology's focus on the social rather than biological roots of human behaviour, and in doing so provide further insights into the 'human' and 'other animal' division in sociology.

Sociobiology

Edward O. Wilson defines sociobiology as: 'The extension of population biology and evolutionary theory to social organisation' (1978, p. x). Sociobiological thinking has been highly contentious in sociology because 'hard-headed sociobiologists' (to use John Dupre's words) see humans as one species of animal like any other and 'just like any other species, they have a set of genetically determined dispositions to behavior' (Dupre, 2004, p. 892). So, sociobiologists propose that there is such a thing as 'human nature' that enables and constrains human behaviour. Charles Darwin's (1998) evolutionary theory is essential to sociobiological thinking because human nature, like the nature of other animals, is understood to be the outcome of natural selection. For sociobiologists, inherited behaviours are those that have enabled species to survive, and it is the behaviour of the human species that has led to human superiority. Nevertheless, this is not to say that sociobiologists see humans as all the same; rather sociobiologists claim that there are natural differences among humans in terms of, for example, their propensity towards the caring role and their degrees of intelligence. Sociobiology thus disputes sociology's implicit assumption that humans are born the same and that it is society that produces differences among humans (e.g. see Guo (2006)). Nevertheless, sociobiologists also observe behavioural similarities among humans (an often repeated example is parents protecting their children) which, they claim, are instinctual. It is the instinctual behaviour that is credited with the survival (and superiority) of the human species. Sociobiology is a diverse field and there are many differences among sociobiologists, for example, in the credit they give to the biological determinants of human behaviour. For this reason, as Dupre makes clear, sociobiologists are often puzzled by the criticisms of their work because, they argue, 'they recognize many other forces, including environmental conditions that interact with the alleged biological determinants of behavior, apart from the purely biological' (2002, p. 149). Be that as it may, sociobiology as an area of study is often accused of biological determinism and even of genetic determinism (Dupre, 2002, p. 149).

We can see why the sociobiological position clashes with the perspectives found in sociology, because sociobiology locates the primary driver of human behaviour in biology, that is in instinct rather than in culture. Such a view of human behaviour is generally rejected in sociology textbooks. As we saw in Chapter 1, Alger and Alger point to what they call distancing concepts, such as 'instinct', that are used in textbooks to distinguish humans from other animals (2003, p. 81). For instance, Anthony Giddens and Mitchell Duneier claim that 'Most biologists and sociologists agree that human beings do not possess instincts' (2000, p. 51). It is arguable whether most biologists do agree with sociologists but it certainly seems, on the basis of what we have seen, that sociobiologists generally do not agree with sociologists. This is most obviously revealed in sociobiological explanations of social inequalities that centre on their 'biological foundations'. Although Wilson (1978) was eager to point out that he was merely *describing* biological differences that he felt lead to inequalities, it is hard to ignore the *justifications* that sociobiology provides for a range of human inequalities. Such biologically deterministic explanations of inequalities among humans have not gone unchallenged from within and outside sociology. A brief examination of assumptions about the biological determinants of gender inequalities provides a good example. In her discussion of the nineteenth century biologically determinist assumption that brain size is a sign of intelligence, Jennifer Mather Saul (2003, pp. 233–4) points to the historical ideas that deemed women to be less intelligent than men because their heads are often smaller than men's. In referring to Cynthia Russett's (1989) work, Saul reveals that many male scientists remained committed to this conclusion even when they were left scratching their comparatively minuscule heads over the observation that elephants have much larger heads than did they (2003, pp. 233–4). This, for feminist sociologists like Saul, reveals a gendered way of looking at the world. Paradoxically, sociology has not traditionally made similar challenges to sociobiological assumptions about 'human' and 'other animal' differences and inequalities; indeed sociology has conventionally buttressed such assumptions (discussed in Chapter 3). Consequently, when biological determinism is applied to humans it is usually accepted by sociobiology but is contentious and is often refuted by sociology. However, when biological determinism is applied to other animals it is usually seen as non contentious by sociology, sociobiology and many other disciplines.

In what remains of this chapter I consider sociological (and at times, other) challenges to biologically essentialist notions of differences among humans to critique notions about (usually ranked) differences between humans and other animals. I also examine how work in the field of genetics has at one and the same time contested and cemented some assumptions about biological difference and similarity.

The discussion enables consideration of the social construction of 'human' and 'animal'; the problem of knowing the mind of another; and challenges to the view that other animals do not have selves. Firstly I examine how sociology has conceptualized 'human'. So far we have seen that sociology is concerned with social life and it is this, or elements of this, that sociologists focus on when explicating the differences between humans and other animals. We have observed that the sociologist Mead considered the human self to be fundamental to the distinction between humans and other animals. In the next section I examine the work of the classical sociologist Emile Durkheim, in order to explore further sociological arguments about what makes humans different from animals.

Human and animal difference: Some arguments from Emile Durkheim

The French sociologist Emile Durkheim claimed that it is culture and society that separates humans from other animals. He maintained that

> ...it is civilization that has made man into what he is; it is this that distinguishes him from the animal. Man is only man because he is civilized. To look for the causes and conditions on which civilization depends is therefore to look, as well, for the causes and conditions of what, in man, is most specifically human. This is how sociology, while drawing on psychology, which it cannot do without, brings to this, in a just return, a contribution that equals and exceeds in importance the services it receives from it. It is only through historical analysis that it is possible to understand what man is formed of: for it is only in the course of history that he has taken form.
>
> (Durkheim, 2005, p. 35)

Echoing Descartes, for Durkheim civilization depends on the human ability to reason, but he does not think that society simply consists in individual human beings; rather he argues that society exists 'sui generis' in that it exists above and beyond the individuals who comprise it. So for Durkheim society has a life of its own (1964 (orig 1885)). He sees this in a range of what he perceives to be observable facts (e.g. that human ideas and beliefs exist beyond the individuals who have them). Taking human notions of what is a 'human' and what is an 'animal' as a case in point, he says, 'The idea of man or of animal is not personal to me; it is to a large extent common to me with all the people that belong to the same social group as me' (Durkheim, 2005, p. 37). Hence, for Durkheim, our notion of 'human' or of 'animal' (or of any other concept for that matter) is common to individuals in society, but it is only because they are common that we can engage in what he calls 'intellectual exchange' with each other (2005, p. 37). What Durkheim is seeking to do is differentiate individual thought from reason, as he claims reason comes from taking society into account. Thus, he argues, 'We possess both the faculty to think under the forms of the individual, which constitute sensibility, and a faculty to think under the forms of the universal and impersonal, which constitute reason' (Durkheim, 2005, p. 41). It is this ability for reason that makes us human.

Durkheim continues with his conception of society as sui generis in his most familiar pronouncement that we should 'treat social facts as things'. By this he means that social phenomena have objective realities that are external to individuals, and these social phenomena have a 'coercive' influence on individuals in society (Durkheim, 1964 (orig 1885)). In this way society has the power to shape what we think and what we do. This is evident in his well-known study of suicide in which he observes that the 'social fact' of regulation (i.e. the rules and directives that govern society) and the 'social fact' of integration (e.g. participation in groups such as family and religion) influence human predispositions to suicide (Durkheim, 1966 (orig 1897)). Durkheim was writing well before the term 'sociobiology' was coined, and it seems clear that his views are very different from the views evident in sociobiology because he sees human behaviour as being constrained and enabled by social facts rather than being founded in biological drives or instincts. Durkheim's sociological study of religious life provides evidence of why the focus on 'social facts' is crucial to his perspective on the difference between humans and other animals.

In his examination of religious life, Durkheim explored the reasons behind the body–soul dualism expounded by Descartes. As we have seen, Descartes argued that not only does the human ability to think distinguish humans from other animals but it is also the basis of human superiority because 'the mind is distinct from and superior to matter' (Sutcliffe, 1968, p. 19). For Descartes, the human mind is evidence of a sacred soul that is different from the corporeal body (1993, p. 281).

Durkheim comments that this 'constitutional duality of human nature' (i.e. this body soul dualism) is 'A belief [that is] so universal and so permanent [that it] cannot be purely illusory' (2005, p. 36). In thinking about such universal concepts and ideas he drew on his notion of the 'collective conscience', which is comprised of the shared beliefs and moral values that exist above and beyond individuals and which binds society together. For Durkheim, the human soul is central to the development of these shared beliefs and moral values because 'Our [human] sensory appetites are necessarily egoistic.... Morality begins only with disinterest, with attachment to something other than ourselves' (2005, p. 36). So morality demands that we move away from thinking about our own needs; and this can only come from detaching ourselves from the 'instincts and inclinations that are the most deeply rooted in our body' (Durkheim, 2005, p. 37). This is something that, in his view, other animals cannot do.

Civilized society demands that humans step back from their selfish desires. In consequence, Durkheim sees human life as a perpetual conflict between, on the one hand, the wants and needs that are driven by our senses and, on the other hand, the moral values that draw us away from concentrating on our individual wants and needs (2005, p. 37). Accordingly, 'There is no moral act that does not imply a sacrifice' (2005, p. 37) and this gives rise to both our 'wretchedness and our grandeur'; wretchedness because we are fated to suffer and grandeur because it is this that distinguishes us from other animals (2005, p. 38). In his words, 'The animal takes its pleasure in a unilateral, exclusive movement: man alone is obliged, as a matter of course, to give suffering a place in his life' (Durkheim, 2005, p. 38). So, for Durkheim, the traditional distinction between body and soul (as found in the work of Descartes) is not imaginary but real (2005, p. 38). Sociology has a crucial role to play in the analysis of this distinction because, he argues, it can answer questions about the source of this duality and antithesis in humans (Durkheim, 2005, p. 38). This brings us back to Durkheim's notion of the collective conscience. Although existing sui generis, the beliefs and values found in the collective conscience must make their way into individual consciousness to ensure the binding together and the survival of societies (Durkheim, 2005, p. 43). In consequence, humans have what Durkheim calls a simultaneous 'double existence': 'one purely individual, which has its roots in our organism, the other social, which is nothing except an extension of society' (2005, p. 44). For Durkheim, this is important for two reasons. Firstly, as we have seen, this is central to his picture of the distinction between humans and other animals (2005, p. 44); and secondly, these 'sui generis virtues' are not mysterious, rather they are 'scientifically analyzable' via sociology because they are perceptible in 'material objects, things, beings of all sorts, shapes, movements, sounds, words, etc. which outwardly sign and symbolize them' (2005, p. 42). Because 'Man is only man because he is civilized' (2005, p. 35) and civilization is established in humans in society, other animals cannot be included in the subject matter of sociology. Although sociobiology sought to challenge these perceptions of the roots of human behaviour, it is more current theories associated with genetics that have exercised sociologists more recently.

'Human' and 'animal' similarity: Some arguments from genetics

Research in human genetics has endorsed the idea that there is a biological foundation to human behaviour. For example, the sequencing of the human genome has been credited with assisting in an understanding of 'the biology of addiction' by enabling researchers to 'identify genes that contribute to individual risk for addiction and those through which drugs cause addiction' (Nestler, 2001, p. 834) and, along with environmental factors, human genetics are seen as having a key role 'in determining aggressive tendencies' (Craig, 2007, p. 227). Because such research promotes the gene as the ultimate engineer in the body, Tim Newton notes that 'genomics has the potential to erode, if not collapse, some of the temporal distinctions between the biological and the social domain' (2007, p. 100). In consequence, Sheila Jasanoff (2005) observes, such research seems to demand 'a fundamental rethinking of the identity of the human self and its place in larger natural, social and political orders' (cited in Newton, 2007, p. 100). Of course, the popular perception of the charting of the human genetic map goes way beyond the actual achievement (Newton, 2007, p. 105) but, nevertheless, genome profiling is being heralded as something that will pave the way for realizing the effective treatment of genetically based human health problems (Turner, 2007). But, genetic research into human behaviour indicates that there is more to this research than the promise of a human health paradise.

Because gene theorists locate (at least some) human behaviour in human biology they are suggesting that we are not necessarily as different to other animals as we once thought. Studies in genetics suggest that, rather than being distinctly different, all species (including human) share over 90 per cent of their genes (Fuller, 2006). In consequence, Fuller (2006) observes that the 'human/animal dualism' and differences among species are convenient labels and schemas rather than natural types. Some sociologists have embraced claims made by gene theorists and in doing so maintain that genetic make-up as well as the environment influence human behaviour (e.g. Owen, 2006, p. 917). This is a long running debate, but what I want to pick up on here is Fuller's (2006) observation (above) that the perceived natural differences between humans and other animals might be merely expedient labels and schemas. This brings me to social constructionism, a sociological approach to knowledge that has been used to counter the ways in which we think about human difference from other animals.

The social construction of 'animal' and 'human'

The sociological debate thus far has focused on the following assertions: there is something natural and immutable about biological differences between humans and other animals; biology is the basis of the behaviour of other animals whereas human behaviour is grounded in the social; and consequently sociology should confine itself to studying humans in societies. We have already seen that natural scientific thinking about the role of the gene has persuaded some sociologists to review assertion two, that is the role of biology in human behaviour; however, I want to explore challenges to the above assertions that come from inside sociological perspectives, from the perspective of social constructionism. The term 'social construction of reality' was first used by Peter Berger and Thomas Luckmann (1967) to point to how human worlds are socially produced. Social constructionism is founded in symbolic interactionism (Chapter 1), which emphasizes the ways in which reality is understood and negotiated. As we have seen, for social interaction to take place there must be some agreement among the individuals involved (e.g. we could not talk to each other if there were no common meanings for words), nevertheless individuals in social situations perceive things differently. Social constructionism stresses the role of perception by asserting that what is seen as naturally given is in fact socially produced (Berger and Luckmann, 1967). We will explore this perspective in depth in Chapter 3; however, here it is useful to consider how the social constructionist perspective can be used to challenge notions of innate and immutable (biological) differences between humans and other animals.

Social constructionist approaches to 'human–animal' difference focus on how the distinction between humans and other animals and, indeed, our notion of 'human' and of 'animal' are socially produced rather than naturally given. Thus, Donna Haraway (1991) argues, there is no purity to 'human' or to 'other animal'. Arnold Arluke and Clinton Sanders explain that 'To say that animals are social constructions means that we have to look beyond what is regarded as innate in animals – beyond their physical appearance, observable behavior, and cognitive abilities – in order to understand how humans will think about and interact with them' (1996, p. 9). Arluke and Sanders seek to counter the idea, found in conventional sociology and beyond, that other animals are purely biological instinctual entities that are essentially different from humans because ' "Being" an animal in modern societies may be less a matter of biology than it is an issue of human culture and consciousness' (1996, p. 9). 'Difference' is clearly important in sociological understandings of human culture. For example, sociological challenges to sociobiological views that suggest an instinctual root to human similarities point to the ways in which notions of 'similarities' overlook a wealth of variations, not least in the ways in which humans assign different meanings to phenomena. Social constructionists argue that sociologists should go further than merely challenging ideas about the biological foundations of human behaviour because phenomena that are seen as naturally given are in fact social constructions. From this perspective, the human view of what is 'natural' and what is 'societal' is itself socially produced. As Arluke and Sanders make clear, social constructions manipulate what we think because 'they do nothing less than shape our consciousness' (1996, p. 16). This is plain when we consider how the meaning of 'animal' seems to be naturally given, permanent and unalterable (1996, p. 10). Our notions of 'wild' animals and 'pet' animals appear to be based in the 'natural' make-up of the other animals concerned. So, tigers are seen as 'wild' because they are regarded as ferocious and naturally untameable whereas bulldogs are seen as 'pets' who are innately gentle and tameable. However, as Arluke and Sanders note, 'What the human consciousness takes for granted as innate biology is, at least in the case of the bulldog, the result of decades of social (and genetic) construction' (1996, p. 18). So, Keith Tester argues, other animals are 'blank paper' that can be inscribed with 'any message and symbolic meaning, that the social wishes' (1991, p. 47). For this reason, argue social constructionists writing in this field, sociology should focus energy on investigating the constructions of other animals. I will explore this in later chapters.

Animal selves

In Chapter 1, we saw that Weber thought that sociology should examine how human meanings and actions shape society. He was interested in the ways in which individuals behave, the subjective meanings that individuals attach to their behaviour, and how their behaviour takes into account the views and behaviour of others (Weber, 1978 (orig 1925), p. 4). Because many sociologists (e.g. Mead) assume that other animals do not have selves (because they see them as purely biological entities), it follows that Weber's approach to sociology could not be useful in the study of other animals themselves. Even sociologists who might be interested in undertaking sociological study of the behaviour of, and interactions between, other animals and humans might feel deterred from trying to make sense of what other animals think and do. This brings me to the problem of other minds, a philosophical position that 'places the burden of proof on those who believe that we can make sense of those with whom we do not have direct contact' (Fuller, 2006, p. 84). Humans are unable to actually communicate with other animals, thus we have made assumptions about, for example, what they feel, whether they have selves and whether they anticipate the future. Our assumptions about them often denigrate them (and lead us repeatedly to treat them appallingly) and we take these assumptions to be realities. Nevertheless, however grounded in reality we think our assumptions are, the burden of proof is on us. As we have seen, a central assumption in sociology is that other animals do not have selves. In offering a critique of this assumption the sociologist Leslie Irvine has sought to examine the behaviour of other animals. She asks two questions, do other animals have selves and is this relevant to sociology (Irvine, 2007, p. 5). In order to gain answers to these questions Irvine turns to research that explores the behaviour of other animals. She starts by developing a 'working definition' of the self, which centres on the self as 'an image (or images) of ourselves (as an object) that appears in consciousness, around which we adapt our behavior' (Irvine, 2007, p. 7). Contrary to suggestions that other animals do not display such consciousness, Irvine argues that there is a good deal of evidence that shows that many species adapt their behaviour according to expectations. For example, she points to Clinton Sanders' (1999) observations of the training of dogs, which show that dogs modify their behaviour according to human expectations. Sanders' work reveals that dogs engage in symbolic interaction (Chapter 1) in that they define situations, adapt their behaviour and select courses of action. Observations of interactions between cats and humans reveal similar behaviours. Alger and Alger's research shows that the behaviour of cats is 'strongly linked to social goals' in that they seek out affection, and 'engage in greeting rituals' (1997, p. 79). Thus, they conclude, symbolic interaction is 'a widely distributed ability throughout the animal kingdom enabling animals to survive more effectively in a large variety of environments' (1997, pp. 79–80). Irvine augments her critique by taking to task sociologists who argue that other animals do not use language. She points to research that shows that animals have 'referential communication', for example birds and monkeys use alarm calls that signal danger from predators, and the recipients of the calls behave accordingly (Irvine, 2007, p. 8). In addition to these symbolic interactions, the selves of other animals are evident in the actuality that events in the lives of other animals shape who they are. Kenneth Shapiro (1989) observes that:

> History informs the experience of a particular animal whether or not it can tell that history. Events in the life of an animal shape him or her... [My dog] is an individual in that he is both subject to and subject of 'true historical particulars'... I cannot replace him, nor, ethically, can I 'sacrifice' him for he is a unique individual.
> (in Arluke and Sanders, 1996, p. 26)

Shapiro's observations show that other animals enrich our lives and our sense of who we are. In order to do so, Irvine argues, they must have selves as we interact with them as individuals (2004, p. 116). For Irvine, such observations and research show that other animals are complex creatures that, like humans, are more than merely biological entities.

Concluding remarks

In examining the ideas behind biological deterministic accounts of the differences between humans and other animals, in this chapter I have explored assumptions that have been used to explain or even justify the relative lack of sociological enquiry about other animals. Descartes' views are fundamental to such assumptions and, although few would now agree that other animals are like machines, still other animals are considered by many to be ruled by instinct and biology and, consequently, as not relevant to the study of sociology. Although sociology has continued to challenge biological deterministic accounts of human differences, in the main sociology has been silent about, or has colluded with, biologically deterministic explanations for the 'human' and 'animal' distinction. Taking human behaviour out of the sphere of biology (i.e. from nature to nurture) has been an important and worthy hallmark of sociology as this has been used to challenge essentialist and discriminatory ideas about human differences. But sociology has generally failed to apply this critical approach to the human exploitation of other animals. We explore such inequalities in the next chapter.

3
Animals, Social Inequalities and Oppression

Introduction

In sociology, the examination of inequalities and oppression centres on the power relations, inequalities and disadvantages that are fundamental to most social systems. Apart from being central to relations among humans such inequalities and oppressions are fundamental to relations between humans and other animals. Despite the prevalence of inequalities associated with being 'animal', sociological analysis has largely centred on power relations among humans. In this chapter I aim to explore how other animals have been (and can be further) drawn into sociological analysis of social inequalities and oppression. In order to do this I examine sociological work that considers human relationships with other animals and the contradictory ways that humans view other species, and investigate the ways in which the oppression of other animals is interconnected with oppressions related to, for example, gender, class and 'race'. In addition, I draw on sociological thinking about difference, similarity and 'otherness' as central aspects of stratification and oppression. The chapter is organized around David Nibert's (2002) theoretical framework, which sets out to explain the origins of the oppression of humans and other animals. To put this theoretical framework in context the chapter begins with discussion of how stratification has been discussed in sociology.

Social divisions and inequalities

Discussions about social divisions, inequalities and oppression in human societies often use vocabulary associated with, on the one hand, 'stratification' systems and, on the other, 'class' systems. Pierre Bourdieu argues that the antagonism between theories that describe social divisions in terms of 'stratification' and those that talk about 'class' relates to divergent perceptions of the world, that is either in terms of consensus (stratification) or in terms of conflict (class) (1984, p. 245). Although a scrutiny of the sociological literature reveals that this is demonstrably true, it is also possible to find literature that does not employ such a rigorous approach, because, at times, the terms 'class' and 'stratification' are used interchangeably (e.g. see Edward Shils, 1962, p. 249). While the difference between conflict and consensus is important in our discussion, the ways in which the terms 'class' and 'stratification' have been used by sociologists need not detain us for too long. What I am interested in here is how social divisions and inequalities are structured in hierarchical ways, and how we could use the terms stratification or class to describe these.

Societies are usually conceptualized as hierarchically organized structures based on social inequalities among humans associated with, for example, differences in age, economic class, ethnic origin, 'race', gender and sexuality. Sociologists differ in the ways in which they see stratification and class systems working. Structural functionalist theorists, such as Talcott Parsons, are concerned with how elements of a society function for the working of the whole society. Parsons defined 'social stratification' as 'the differential ranking of the human individuals who compose a given social system and their treatment as superior and inferior relative to one another in certain socially important respects' (1940, p. 841). For Parsons, social divisions are based on consensus as stratification is a 'functional prerequisite of the social system' in that a social system will only work if individuals are motivated to fulfil required roles (e.g. see Calhoun et al., 2002). Kingsley Davis and Wilbert Moore concur as they argue that social inequality is an apparatus that has 'unconsciously' emerged in societies to 'ensure that the most important positions are conscientiously filled by the most qualified persons' (1945, p. 243). Such a notion of stratification overlooks the often gross inequalities that are associated with stratification systems. In this regard, Karl Marx and Friedrich Engels argued that society is characterized by a split 'into two great classes ... Bourgeoisie and Proletariat' (2009 (orig 1848), p. 35). These classes are in conflict with each other as the bourgeoisie owns the places of paid work and the equipment within, employs the proletariat (the working class) and makes a profit out of their labour; and the proletariat is comprised of those 'who live only so long as they find work, and who find work only so long as their labour increases capital' (Calhoun et al., 2002, pp. 80–1). So, for Marx and Engels, inequalities are based on economic relations, because the bourgeoisie has the political power and proletarians are merely 'a commodity'.

Weber agrees that social structures are based on divisions and inequalities; however he argues that inequalities in power are not determined by economic relations alone. Weber (1978 (orig 1925)) classifies inequalities in terms of 'class', 'status' and 'party'. When talking about a 'class situation' Weber is referring to an individual's economic position in society. Class situation can influence 'status' but does not necessarily do so as status refers to the social estimation of honour which is associated with, for example, lifestyle choices. Thus, Weber argues that status 'is a quality of social honour or a lack of it and is in the main conditioned as well as expressed through a specific style of life' (1978, p. 932). So, a major difference between status and class is that status is usually related to consumption whereas class is associated with production. Weber places classes within the 'economic order', status within the 'social order' and in addition to these he argues that ' "Parties" reside in the sphere of power' (1978 (orig 1925), p. 938). Parties are usually powerful and influential groupings, most obviously political parties, that influence social action. For Weber, parties may or may not be linked to class or status. Although Weber puts forward a more complex analysis than did Marx and Engels, still Weber (1978 (orig 1925)) centres on conflict. He sees conflict as an integral part of society because there are always struggles between individuals, corporations, employers and employees, political parties, and so on. However, power is distributed unequally so each side in any conflict will not be in an equivalent position to win. Accordingly inequalities are endemic to society. A more recent theorist, Ulrich Beck, perceives a 'process of individualization' occurring in developed countries, which is based in the changing 'social meanings of social inequality' (1992, p. 92). He observes that individuals have become individualized in that they are becoming the central unit of social life and, as a result, individuals are less attached to groupings such as 'social classes'. In consequence, social inequalities are seen as the result of individual inadequacies (Beck, 1992, p. 100).

An important feature of the theories referred to above is the different emphasis that each places on inequality and oppression. Although inequality (in its many forms, including inequality in wealth and inequality in rights) is a central feature of analyses of 'class' and of 'stratification', structural functionalists see (at least some) inequality as being a positive and a necessary organizing feature of society whereas Marx and Engels, Weber and Beck, in different ways, see a negative dynamic based on exclusion, exploitation and oppression. Although the above theories differ there is at least one element in common to them all (and in most other theories about the structure of inequalities), that is they centre on inequalities and oppressions among humans. Such a focus excludes a major defining inequality and oppression, that is between humans and other animals.

Speciesism and oppression

As we saw in Chapter 2, in sociology (like in most other areas of thought) biological classifications of 'species' are used to make a hard distinction between humans and other animals. This biological classification of humans on the one hand and 'animals' on the other is not simply used to refer to difference, it is used to stratify or rank order the distinction, with human (usually) perceived to be superior to 'animal'. Sociology has typically opposed biological classifications that have been used to explain inequalities among humans (e.g. related to gender and 'race') but has done relatively little to challenge biological explanations for inequalities between other animals and humans. Indeed, sociological analysis has often served to entrench such inequalities by either ignoring them or by explaining them away as natural. By looking through the lens of inequalities among humans and at oppressions such as sexism, racism and classism, in this section I want to explore speciesism as a way of explaining the acceptance of such species inequalities. Speciesism is a perspective that allows the interests of those of one species to override the greater interests of members of other species (Ryder, 1983 [1975]; Singer, 1990). Many activists and scholars have compared speciesism to racism and sexism (Nibert, 2002, p. 6), but unlike racism and sexism, speciesism has been little recognized, let alone rebuked, by or in sociology. This is, to say the least, a disappointment because looking at speciesism as one of a number of oppressions provides additional perspectives on the oppression of groups of humans and of other animals. Nibert too is disappointed as he observes that sociological perspectives on the oppression of other animals can help us to understand the origins of speciesism and can assist in an understanding of the relationship between speciesism and the oppression of 'devalued groups of humans' (2002, p. 6). The societal implications are obvious when speciesism is recognized as a phenomenon that is over and above an individual affectation. Nibert takes speciesism to be more than merely prejudice because prejudice, he contends, is 'an individual predisposition to devalue a group of others' (2002, p. 8). Rather, Nibert sees speciesism as an ideology which, like other –isms, works as 'a set of socially shared beliefs that legitimates an existing or desired social order' (2002, p. 8). Like racism, sexism and classism (to give three examples), speciesism has social structural and economic causes, it is institutionally based, it supports oppressive social arrangements (Nibert, 2002, p. 10) and it 'legitimates and inspires prejudice and discrimination' (Nibert, 2002, p. 17). In making this contention, Nibert seeks to draw on sociological theory to provide insights into the causes of oppression.

Drawing on Donald L. Noel's (1972) 'theory of ethnic stratification', Nibert puts forward a 'theory of oppressions' in which he identifies three interacting factors that, he argues, are central to the growth and continuation of the oppression of humans and other animals. These factors are 'economic exploitation/competition'; 'unequal power, largely vested in control of the state'; and 'ideological control' (2002, p. 13). The first factor (economic exploitation/competition) points to the ways in which humans and other animals are used as economic resources for the benefit of a minority of humans. Factor two centres on the unequal distribution of power in society and on the ways in which one group, powerful humans, can exert power and control over other humans and other animals. The third factor draws attention to how particular groups (certain groups of humans and all other animals) are devalued by powerful groups. These factors are not discrete; they interact with each other. In the rest of this chapter I organize my discussion around Nibert's three factors as this enables consideration of social inequalities as they relate to other animals.

Economic exploitation/competition

The economic exploitation of other animals is plain to see. Humans and other animals compete for land and over other resources, and this competition often serves the economic ends of powerful humans; it is other animals and less powerful humans who usually lose out. The sociologist Ted Benton (1993, pp. 62–6) discerns nine broad types of human relationship with other animals, which serve to highlight a range of issues associated with the economic exploitation of other animals by humans. It is worth exploring each of these in turn. Benton starts off with the relationship in which humans use other animals to 'replace or augment human labour' (1993, p. 62). This relationship centres on, for example, the use of the bodily power of other animals to shift and carry heavy weights and the utilization of the special sensual abilities of other animals (e.g. smell or hearing) to aid humans. Benton's second type of relationship focuses on the use of other animals 'to meet human bodily needs', most evident in the use of other animals for food, clothing and as 'models' in biomedical experiments (1993, p. 62). Benton's third relationship spotlights the human use of other animals as a source of entertainment (1993, p. 62). This covers a range of events and activities including visiting zoos, watching horse-racing and dog-racing, participating in 'hunting' and killing, betting on dog fights, and viewing a range of popular and fine arts (1993, p. 62). Benton talks about the 'edificatory' use of other animals in his fourth type of relationship (1993, p. 63) because here he centres on the ways in which humans try to learn from other animals. He admits some overlap here with his other relationship types as visits to zoos and engaging in experiments on other animals are just some examples of ways in which other animals are utilized by humans for educative purposes. In the four relationship types above, other animals are often constructed as the private property of humans, who can be bought and sold and treated as sources of profit. The use of other animals as sources of profit comprises Benton's fifth relationship type, and in this relationship category he also refers to specific practices such as industrial farming and the testing of cosmetics and other such goods on other animals (1993, p. 64). Benton stresses that intensive farming and safety testing are especially important in this type of relationship as these activities 'arise as a direct consequence of commercial relations and pressures' (1993, p. 63). In his sixth relationship type, Benton extends his thinking about private property by reflecting on how other animals are used for the physical protection of property, for example when they are used as guard dogs or as police horses (1993, p. 64). Other animals, such as cats and dogs, have become companions to humans, and being treated as 'pets' they play important roles in the personal lives of humans. This relationship comprises Benton's seventh category (1993, p. 65). Category eight centres on the ways in which other animals are used as symbols and metaphors and here we might think about other animals as objects of worship and as symbolic of human emotions (1993, p. 65). Benton's final category focuses on the refutation of a human relationship with other animals such as in the case of the notion of the 'wildness' of other animals who are said to live outside of any relations with humans (1993, p. 65). Benton's categories of relations serve to focus our attention on the countless ways in which other animals are used as sources of profit (1993, p. 63). This is extremely important as, Nibert notes, the ever spiralling drive for profits results in the exploitation of humans as well as other animals; it is generally those humans who are affluent and powerful who benefit from the exploitation of others (2002, p. 94). In consequence, the oppression of other animals is co-dependent on the oppression of humans with 'each practice of exploitation enabling the other' (Nibert, 2002, p. 95). The perception as 'other' is central to this process and practice because other animals are seen as 'other' than human, and devalued humans are seen as 'other' than valued humans. It is this construction as 'other' that permits their economic exploitation (Nibert, 2002, p. 13), and this construction is based on relations of unequal power, which brings me to the second of Nibert's three factors.

Unequal power

Nibert's second factor points to the central role that power plays in the oppression of other animals (2002, p. 13). Power is a crucial and complex concept in sociology, though it has been reserved mainly for unequal relations among humans. Giddens (1987) refers to the 'transformative capacity' of power that enables (some) humans to change events and to change their social worlds by intervening in what is going on. This transformative potential is important when thinking about the goals of social movements in Chapter 9, however here we centre on the negative capacity of power which is associated with inequalities and oppressions. The negative capacity has gained the most scrutiny from sociology. For example, in his analysis of power Mills defines power as being determined by who is and who is not involved in making decisions, and in contemporary societies it is rich and powerful elites that dominate (1970, p. 50). Weber (1978 (orig 1925)) and Nibert (2002) centre on the ways in which powerful actors or groups can exert their will against others in spite of resistance. Nibert emphasizes that humans have developed numerous ways of abusing their power against groups of humans and other animals, in the forceful use of weapons, torture, killing, control, capture and enslavement (2002, p. 13). Of course, in order to successfully maintain its domination, the oppressing group must have the power to subordinate the oppressed group, and it is those who are defined as 'other' who tend to be exploited most easily (Nibert, 2002, p. 13). The subordination that comes from oppressive power can be realized through force or through ideological control (Nibert, 2002, p. 13) as discussed below.

Ideological control

For oppression to be legitimized it must seem to be appropriate to those inside and outside the oppressing group (Nibert, 2002, p. 13), and such legitimization is founded in ideological control. Ideologies are beliefs that legitimate powerful interests and thus justify the subordination of one group by another and, as such, ideologies that devalue a whole group (such as racism, sexism, classism and speciesism) are socially constructed (Nibert, 2002, p. 13). These socially constructed notions serve to naturalize and justify social stratification/class systems. As we have seen, an element of this oppression is the social designation of a group as 'other', and Jacques Derrida's (1982) ideas throw light on how this works. Power and decision-making are central to Derrida's discussion of the identification as 'other' and this is bound with the notion of 'identity'. Drawing on Derrida's work, Ernesto Laclau argues that it is through the exclusion of the 'other' that we can see that the formation of identity is an act of power as it is 'us' who have the power to make decisions about whom to exclude from our definition of 'us' (1990, p. 33). So, those with power (e.g. 'humans') have the power to define who is not 'human' (e.g. 'animals'). But, Derrida adds, the formation of identity is not just a matter of excluding those who are defined as 'other', rather the exclusion is established in hierarchically defined binary oppositions (such as black/white, man/woman, human/animal, person/thing) (Laclau, 1990, p. 33). Within these binary oppositions one category is defined as superior to the other; the excluded second category, what Derrida calls the 'marked' category (e.g. black, woman, animal, thing), is defined as subordinate or lower (Laclau, 1990, p. 33). So, thinking about the identities 'human' and 'animal', the category 'animal' is defined as subordinate in society (and in sociology), and as we saw in Chapter 2, this is established in ideas about 'natural' biological distinctions between 'them' and 'us'.

This brings me to the concept of 'identity'. Identity is a familiar feature of our everyday lives and is central to our idea of who we are. For this reason identity is a central concept in sociology. Recent sociological theories about identity centre on individual identity as changeable rather than permanent. Thus, Stuart Hall (1996) maintains that identities are 'fragmented' rather than 'fixed'. Humans do not have one identity that is the same in all situations and for all time, rather each of us changes on a day-to-day basis depending on, for example, where we are and who we are with. From this we can see that one strand of the sociological discussion about identity focuses on the individual, and in this the concept of the 'self' is crucial. Mead (1934) argued that we develop our sense of self through interaction with others (Chapter 1). Consequently, the self cannot exist without society and, because we see ourselves as reflected by others, we often change our 'self' accordingly.

So, as Giddens argues, 'individuals tend to develop multiple selves in which there is no inner core of self identity' since we manipulate our conduct and our appearance in order to fit in with particular requirements and locales (1991, p. 100). However, although our identities might change (and I will look at this below), what these sociologists also make clear is that a sense of belonging with others is vital in identification (i.e. how we identify with others), and belonging is an important feature of stratification systems (both among humans and, as we will see, between humans and other animals).

Belonging centres on the individual within the collective group or society. This focus on the collective is another strand of the sociological discussion about identity. Richard Jenkins contends that the most significant difference between the individual and the collective is that individual identity emphasizes difference and collective identification emphasizes similarity (2004, p. 16). Accordingly, we might see our individual identity as something that is unique to us whereas collective identity refers to the sense of belonging we have with a larger grouping. Collective identities might be based on, for example, gender, class, ethnic origin or culture, and such collectivities provide frameworks of resemblance to and difference from other people (Jenkins, 2004; Peggs, 2009a). Hence, both resemblance and difference are fundamental to the ways in which we think about identity because, as Hall points out, we identify with some through the capacity to exclude others (1996, p. 5). So, my (individual) identity as a woman relies on the exclusion of men from the (collective) identification of women. However, in his discussion of identity and identification Jenkins is focusing on humans alone as he defines society as 'the human world' (2004, p. 16); we can broaden this out. For example, coming back to my identity as a 'woman' (i.e. a *human* female), this identity relies on the exclusion of all males, all human females under a certain age (who might be identified as 'girls') and all female other animals. As I have discussed elsewhere (Peggs, 2009a, p. 87), this shows that difference and similarity are muddled concepts, because in order to achieve similarity with some we have to disregard a number of the differences among us (and vice versa). The example I give elsewhere centres on what we do in order to identify ourselves as 'human' rather than as 'animal' (Peggs, 2009a, p. 87). Identification as human can only be attained though the recognition of 'significant' similarities (e.g. human language use) and the discounting of differences judged to be less 'significant' (e.g. density of body hair). Additionally, differences among those designated 'animal' (e.g. ability or not to fly unaided) must be overlooked in order to conceive all other animals as different from 'human' (Peggs, 2009a, p. 87). It is this exclusion of 'the other', as Nibert (2002) makes clear, that is crucial when thinking about stratified differences.

The collective identity 'human' is founded in our notions of our 'natural' uniqueness, a uniqueness that we associate with human superiority (Peggs, 2009a, 2009b), but, as we saw in Chapter 2, the grounds for our ideas about our natural uniqueness are becoming increasingly unsettled. For instance, because all animals share most of their genes, it seems that species are not natural types but instead are convenient classifications (Fuller, 2006, p. 29). Furthermore, there is no distinct and even boundary between humans and other animals as some other animals are considered to be closer to humans than are others. For example, great apes such as gorillas and orang-utans are often viewed as 'honorary humans' (Midgley, 2004, p. 147) and thus are considered to be superior to many other animals in stratified terms. Accordingly, within the group 'other animal' (or 'animal' as the group is conventionally labelled) there is a complex stratification system that differentiates on grounds of whether the other animal is vertebrate or invertebrate, mammal or non-mammal, large or small, endangered or not, 'domesticated' or 'wild' and so on, and this system complicates the simple 'human–animal' distinction. This complexity leads Derrida to conclude that '[t]here is no animal in the general singular, separated from man by a single indivisible limit' (2004, p. 125). So it seems clear that the category 'animal' is simply a convenient all-encompassing label based on assumptions about the 'natural' shared characteristics of this designated group. This brings me back to social constructionism. Social constructionist perspectives allow us to explore the ways in which human and other animal identities are products of human decision-making, which leads me to ask the question why do we take assumptions (such as those that suggest that there are natural hierarchical distinctions between humans and other animals) to be true? This brings me to the role of discourse.

Human language is a central feature of society. It is human language that enables us to communicate with each other, and sociologists such as Mead propose that humans are unique in their ability to develop and use language (see Chapter 2). Certainly, this line of thinking is contested. For example, Michael Agamben argues that other animals have language, though he retains the break between human and other animals (see Calarco, 2008, pp. 84–5). Nevertheless, the use of language is a central aspect of the foundation of human group identity, and sociology has explored how individual or group identity among humans is established by the language that they use (e.g. Fishman,

1972). Human language could be taken as simply a way of describing the world; however, as Michel Foucault (1980) makes clear, language is a crucial feature of social power. In this regard Foucault refers to language as 'discourse', that is language as associated with a particular group of ideas. A good example can be found in discourses used about various animal rights movements, describing them either as animal 'freedom fighters' or 'terrorists' (we explore this in detail in Chapter 8). We get a very different picture of those involved depending on which description is used. What this shows is that discourses *describe* and *define* the world because discourses affect our views on all things. So, it is not possible to avoid discourse, and nothing exists outside of discourse because everything is constructed from within discourses (Foucault, 1980, p. 118). Therefore, what are taken to be truths are actually socially constructed from within discourses (Foucault, 1980, p. 118); we cannot get to the 'truth' out there, outside of our way of describing the world. Returning to the purported naturally stratified distinction between humans and other animals, analysis of discourse suggests that the ways in which we communicate about this division actually constructs this division. The distinct discourses used about various animals, for example describing them as either 'animal' or 'human', construct how we view ourselves and the other animals under discussion. Humans generally view other animals to be different and lower, and this difference is constructed in a discourse, that is, to use Foucault's words, 'neither true nor false' (Foucault, 1980, p. 118). A social constructionist critique suggests that our notions of 'animal' and 'human' are purely ideas, not truths; they are ideas that are constructed within discourses and are maintained through the use of discourses.

I have explored the construction of 'animal' and 'human' identities in my work on the ethics of experiments on other animals (e.g. Peggs, 2009a). In this work I am especially interested in exploring the role of the construction of identities in the oppression of other animals. As we have seen, sociology has centred on 'fragmented' rather than 'fixed' identities (Hall, 1996); however, some identities appear to be fixed, or at least more fixed than others. In this regard, Jenkins refers to 'primary identities', which are established in earlier life and are 'more robust and resilient to change later in life than other identities' (2004, p. 19). He offers gender identity as an example. Gender is established at birth and is usually seen as fixed. I have used this example (in my work 2009a) to explore humanness as an identity. In this discussion I turned to the work of Judith Butler, which enabled an exploration of how identities such as gender seem more robust, in the sense that they are so ingrained that we feel them to be natural even though they are not. Butler's (1999) notion of the 'performative subject' as it applies to gender identity provides an effective example. In declaring that gender is 'performative', Butler (1999) opposes the idea that there is something natural to gender identity. Butler is not saying that organic differences do not exist, what she is saying is that 'what we take to be an internal essence of gender is manufactured through a sustained set of acts, posited through the gendered stylisation of the body' (1999, p. xv). So, 'gender is not an expression of what one *is*; it is what one *does*' (Lloyd, 2005, p. 25, original emphasis). For example, in saying that my name is Kay rather than Karl I am 'doing' gender as I am doing being a woman rather than doing being a man. It is not only identities *among* humans that are performative but also, as I have argued elsewhere, human identity itself is performative (Peggs, 2009a). So, taking Butler's contention, I argued that human is not something that we are, rather it is something that we do. So, when we communicate through, for example, human language we are 'doing' human. To complicate this further, the discourses we use to talk about 'language' often define 'language' as human; so when we are defining language we are also usually doing human.

Speciesism and interlocking systems of domination

By way of a conclusion to this chapter I want to explore the links between speciesism, racism and patriarchy, by drawing on the work of Carol J. Adams. Adams is clear: 'exploitation entails more than the exploitation of animals' (1995, p. 15). In maintaining that the exploitation of other animals is one of a range of exploitations from which privileged groups benefit, Adams argues that the 'human–animal dualism is embedded within a racist patriarchy' (1995, p. 15). Just as 'race' and gender are social constructs, claims Adams, so is species. However, there is more to her argument than this. What Adams is especially interested in is what the writer bell hooks calls the interlocking systems of domination (1995, p. 79), because 'some people and all animals have been cast as "others"' (1995, p. 78). Thinking about the positions we hold within these systems of domination she notes that we can be sometimes victims of oppression while at other times we can be beneficiaries (1995, p. 81). This is important for two reasons. Firstly, recognition that we can be beneficiaries and victims militates against an 'additive model' of oppression, a model that has also been criticized by theorists such as Patricia Hill Collins. Collins sees distinctive oppressions associated with 'race', gender and class as 'part of one overarching structure of domination' (Adams, 1995, p. 79). Thus, I might be the victim of sexist oppression but be a beneficiary of racist oppression. Secondly, seeing this overarching structure makes room for an analysis of the connections between the oppression of humans and the oppression of other animals. The historical positioning of woman between the two poles of man and of other animals, with woman perceived as closer to the natural (i.e. other animals) and inferior to man, is an instance of such connections (Adams, 1995, p. 79). Another example referred to by Adams is racist discourses that position people of specified ethnic groups as closer to 'animal'. Arluke and Sanders' discussion of the atrocious and murderous treatment of groups of people in Nazi Germany who 'were viewed as "lower animals" to be dispatched accordingly' (1996, p. 161) provides an important illustration of the ways in which the brutal and discriminatory treatment of humans is connected with the brutal and discriminatory treatment of other animals. Moreover, for us to perceive the cruelty and odiousness of this treatment of groups of humans we must recognize that 'lower animals' are treated in this way. Despite these clear examples, sociology has often steered clear of exploring links between the oppressions of groups of humans and the oppressions of other animals and has remained very quiet about the oppressions of other animals themselves. Arluke comments that such omissions might be a symptom of 'political and psychological insecurities' that see sociological enquiry into the oppression of other animals as debasing the study of important human oppressions (2002, p. 1).

Stratification systems that differentiate humans from other animals are a major characteristic of most, if not all, human societies. Although systems vary across the globe, in the main other animals have unequal access to resources and have unequal status among themselves and in comparison with humans. Such speciesism is not an aberration of the present; it has historical precedent. In consequence speciesism often goes unnoticed; it is defined as natural and fair, and thus not as speciesism at all. Speciesism is inextricably linked to the power of the human, and human language is a crucial feature of social power. Discourses associated with the idea that humans are superior to other animals construct how we view ourselves and other animals, and these discourses see the place of humans as being associated with human 'natural' superiority over other animals. Identity is a crucial feature of stratification systems as shared human identity engenders a sense of belonging (to 'us') as against exclusion (of 'the other'). This 'boundary work', (that is 'the drawing and blurring of lines of demarcation between humans and animals' (Arluke and Sanders, 1996, p. 133)), is essential to the interlocking nature of stratification systems. Consideration of how systems of domination interlock makes clear how some groups of people and all other animals are marked 'other'. Connections can be made between oppressions such as racism and sexism and speciesism, but we should not use an additive model, because at one and the same time we can be the victims of oppression and the beneficiaries of oppression. We will explore this throughout the book and will come back to it in more detail in the final chapter. In the next chapter we explore a clear and obvious element of the oppression of other animals: cruelty to and abuse of other animals.

4
Animals, Crime and Abuse

Introduction

One of the most serious charges that can be laid against humans is that they have behaved like 'animals' (Midgley, 2004). This insult is often reserved for the most shocking of human behaviour, behaviour associated with crimes such as rape and murder. Consequently, in the popular media as well as in the news media it is not unusual to find humans who have perpetrated the most awful of crimes being called 'animals'. However, it is not other animals who commit such acts; the perpetrators are human. Indeed, many of the victims of human crimes are other animals. The abuse of other animals is a daily occurrence in many societies and the criminal law authorizes the abuse of other animals (e.g. in laboratories and on farms) in their billions. This chapter explores the abuse of other animals through the lens of the sociology of crime, and in doing so examines the ways in which the perpetrators of crime are 'animalized' (Arluke and Sanders, 1996), the ways in which other animals are at risk from humans (e.g. through neglect and cruelty) and the ways in which humans use other animals for hazardous purposes (e.g. as status symbols and as weapons in gang violence). The chapter also explores the relationships between the abuse of other animals and the abuse of humans, which has been a key focus of crime and other animals. I begin by considering the interactions between morality and crime.

Sociology and crime

'Crime' is an essential feature of the study of sociology because crime and deviance are ubiquitous in societies. Indeed, Durkheim (1947) argued that crime is not only prevalent in diverse societies but necessary to those societies. Thus, for Durkheim (1947), crime is a fundamental condition of all social life not least because it is necessary to the evolution of morality. While at first sight this is a perplexing assertion as we would normally associate crime with immorality, what Durkheim (1947) is referring to is the way in which we need the 'bad' in order to understand the 'good'. Accordingly, Durkheim (1947) argued that crime and deviance perform four positive functions in society. Societies are layered by norms and values that guide human behaviour and the first function of crime and deviance is that they affirm these norms and values. For example, and without getting into the complexity of the Act, the UK Wild Mammals (Protection) Act 1996 protects mammals who live in the wild 'from certain cruel acts'. Offences listed under the Act include the mutilation, kicking, beating or nailing of mammals. These offences correspond with norms and values about the treatment of mammals and also include norms and values about which other animals are not covered (i.e. other animals who live in the wild who are not mammals). The capacity to deviate from these norms and values is associated with freedom, but that freedom has a price, which brings us to the second function. This second role of crime and deviance refers to the way in which the classification of someone as deviant or as a criminal draws a borderline between right and wrong; certain behaviours have largely societal acceptance as crimes and there is an acceptance that deviance from norms and values should be punishable. Responding to deviance not only clarifies moral boundaries but also promotes social unity, which is Durkheim's third function of crime and deviance. People react to serious crimes with collective indignation and this ties people morally to each other. But, Durkheim makes clear that collective indignation should not be seen as an expression of social unity against particular crimes and particular types of deviant behaviour because crime and deviance encourage social change, which is the fourth function. Deviance and actions labelled as crimes breach society's moral boundaries and, at times, such breaches indicate that changes should be made to legislation of other aspects of the status quo. For example, in England in 1811 Lord Erskine was ridiculed during his House of Lords speech because he advocated 'justice' to 'lower animals'; however, just 11 years later, in 1822, the Martins Act afforded protection to cows (Salt, 1980 (orig 1892), p. 7). Writing in 1892 Henry Salt noted that the passing of this Act 'is a memorable date in the history of humane legislation...for the invaluable precedent which has been created' (1980 (orig 1892), p. 7). Moral consideration for some other animals (in this case cows) became enshrined in law, which signalled a change because,

Salt argued, 'if some animals are already included within the pale of protection, why should not more and more be so included in the future? (1980 (orig 1892), p. 8). So, for Durkheim, crime has an important moral function and Salt's example shows that early legislation heralded a move to greater moral consideration for other animals. Such changes do not only signal changes in moral sensibilities they also show that crime and deviance are social constructions that do not have a reality outside of the definitions they are given in a particular society. It is the study of the social construction of crime that is, suggests Piers Beirne (2002), the focus of criminology.

Sociology as a subfield of criminology

Sociology is a major perspective in the interdisciplinary field of criminology (Beirne, 2002). Beirne defines 'criminology' as a 'discourse that investigates the whys, the hows and the whens of the generation and control of the many aspects of social harm – including abuse, exclusion, pain, injury and suffering' (2002, p. 381). In using the term 'discourse' Beirne is seeking to make clear that categories such as 'crime' and 'criminal' have no ontological status, that is they have no standing as reality, rather, in his words, they are 'social constructions that are selectively applied by a network of state and other social control apparatuses to the actions of some members of society and not to those of others' (2002, p. 381). A good illustration of this is labelling theory.

In respect of crime and deviance, labelling theory focuses on how others respond to reproved actions. Let us consider the following. A human squeezes a noxious substance into the eyes of a rabbit. Is this a crime? If the human is a scientist who has a license to carry this out, the procedure is approved (and might even enable the human to gain an award) as it is considered by law to be useful and necessary for human safety. As a result, the action is not labelled a crime. If the human is a member of the public who has no such license for this procedure it is considered a crime and the person could be fined or more. However, it would only be labelled a crime if the action was detected and if the perpetrator was convicted of the crime. What this example shows is that the meaning of crime is highly variable and is subject to definitions, detection and response. Of course, historical time and geographical space complicate matters further as there are considerable differences between societies, not least in terms of which actions are labelled 'crimes'. So, 'crime' and 'criminal' have no ontological status. As we will see below, this is not merely a matter of definitions as 'when people define situations as real they become real in their consequences' (Thomas and Thomas, 1928, pp. 571–2) and the consequences for other animals are also real and powerful in respect of which actions against them, and under what circumstances, they are labelled crimes. Criminology, as Beirne understands it, tries to 'uncover the sources and forms of power and social inequality and their ill effects' (2002, p. 381). One of the ways in which other animals have entered the field of criminology is through 'positivist criminology'.

Positivist criminology

'Positivist criminology' is one among a number of theories of crime used in sociology. The perspective is associated with work that tries to find biological causes for crime, often grounded in biological links between 'human' and 'animal'. The most notorious theorist associated with this perspective is Cesare Lombroso. In 1876 Lombroso, an Italian physician argued that 40 per cent of criminals are born criminals. His theory was that criminals have a distinctive body type and distinctive characteristics (stigmata), which align them with their ape-like ancestors. The stigmata include excessive hairiness, prominent jaw, low forehead, protruding ears, thick skull, relatively long arms and feet with mobile big toes that are capable of grasping (Gould, 1980). Accordingly, Lombroso argued that many criminals are atavistic throwbacks to an earlier stage in human evolution. He saw their savagery as being a consequence of their biological variety, which he saw as a reversion to a 'primitive' or 'subhuman type' of human who is located somewhere previous on the evolutionary scale. Because criminals are biological throwbacks, their behaviour, he argued, will inevitably be out of kilter with the rules and expectations of modern civilized society. So, for Lombroso, many criminals are destined by their physical constitution to break the law. As a result, he tried to make a case for a preventive criminology because he felt that 'society need not wait for the act itself, for physical and social stigmata define the potential criminal. He can be identified, watched and whisked away at the first manifestation of his irrevocable nature' (Gould, 1980). Lombroso saw limited possibilities for such individuals, one being the 'irrevocable detention for life for any recidivist with the telltale stigmata' (Gould, 1980). Lombroso's ideas seem very outdated now as even research published in the early twentieth century demonstrated no such biological characteristics among criminals (e.g. see Goring, 1913). However, this is not the end of suggestions of a biological connection; Lombroso's theories anticipated more recent genetic explanations of crime (for discussion, see Ferguson, 2009) that attempt to establish possible links between genetic make up and increased propensity to commit crime. However, the controversies around this need not detain us here. For our purposes there are four issues that are important. The first is that crime has been associated with specific physical features that are biologically inherited. The second is that these physical features are seen in other animals (most notably apes). Although the notion of this inheritance is now refuted, more recent theories of an association between genetics and criminality imply that the notion that biology affects the behaviour of other animals applies to at least some of the behaviour of some humans. This leads to the third point, which centres on the ways in which these inherited behaviours are pathologized behaviours. Recalling the discussion in Chapter 3 about the designation of 'other' and the boundary work involved, the notion that criminals are biologically different from noncriminal humans cements the construction of hierarchical difference among humans, and between humans and other animals, because it is only 'throwback' humans who are biologically determined to commit crimes. Finally, the belief in inherited biological traits for criminality is rooted in an acceptance of 'crime' and its associated terminologies as having an ontological status. As we have seen, Beirne (2002) refutes this idea as he argues for a social constructionist approach to crime, albeit one that recognizes that such constructions have real effects. The idea that other animals can commit crimes might seem amusing to us now, but this idea provides evidence of the social construction of crime and the effects of changing constructions.

Other animals as criminals

In the past, other animals were sentenced to suffer and die for their crimes (Ritvo, 1987). Harriet Ritvo (1987) notes that, inherent to the earliest laws for which there are records, as criminal subjects other animals had rights as well as responsibilities (1987, p. 1). Other animals had a legal role, which is very different from current thinking about the roles of other animals. For example, before the nineteenth century, in the absence of human witnesses to burglaries, under Germanic jurisprudence, 'dogs, cats, and cocks were permitted.... to testify in court' (Ritvo, 1987, p. 1). In Britain, France and other countries, other animals such as dogs, pigs and monkeys were hanged for crimes. But by the nineteenth century British authorities had stopped sentencing other animals for crimes, a move that was celebrated by many as the triumph of 'refined and humanitarian modern conceptions of justice' over 'gross and brutal medieval conceptions' (Edward Payson Evans quoted in Ritvo, 1987, p. 2). However, this change in sentencing also signalled a transformation in perceptions of other animals, from seeing other animals as morally accountable beings who are responsible for their actions to viewing other animals merely as the property of humans (Ritvo, 1987, p. 2). As a result, human 'owners' became responsible for assessing the danger that others animals might pose to others or to the property of others, and they were required to act appropriately to prevent instances of injury and damage (Ritvo, 1987, p. 2). Ritvo argues that this change reflected 'a fundamental shift in the relationship between humans and their fellow creatures, as a result of which people systematically appropriated power they had previously attributed to animals, and animals became significantly and primarily seen as the objects of human manipulation' (1987, p. 2). An analysis of these changes is significant for sociologists for a number of reasons. As noted above, the change in the legal position of other animals is a good indicator of the social construction of crime and deviance. The changing role that other animals play and the ways in which other animals have entered discourses about crime mean that they are important to sociological analyses of crime. The change in the legal position of other animals is an indicator of changing ideas about the moral worth of other animals and notions of whether other animals have rights. Thinking back to Durkheim's (1947) perception of crime as fundamental to the evolution of morality, perceptions of the position of other animals within the scope of crime and morality are central issues.

Other animals and moral worth

The issue of rights is fundamental to analyses of crime, and the consequences and considerations of the allocation of 'rights' to other animals are essential when thinking about crime and other animals. In Chapter 3 I reflected on bell hooks' conception of interlocking systems of domination and the ways in which some people and all animals have been cast as 'others'. Although humans are subject to a range of discriminations, the moral standing of humans as a group and the denunciation of discrimination against all humans is at the centre of human rights legislation. Although discrimination against humans is rife, in law all humans have rights and all humans are accorded equal moral standing. Most other animals have moral standing, but their moral standing is considered to be lower than that of humans (and there are differences in moral standing among other animals). Although other animals have some moral standing they are not considered to have 'rights' as such; I will return to 'animal rights', and the complexity and nuances around this notion later in the book (Chapter 8). Here I want to draw on Robert Garner's notion of the 'moral orthodoxy' (2005a, p. 15) and how this orthodoxy informs ideas about crime and other animals (especially in terms of cruelty to and abuse of other animals).

Garner (2005a) refers to what he calls a 'continuum of recognition' of the moral standing of other animals. At one end of this continuum is the idea that other animals have no moral status because, for example, they are considered to lack sentiency (e.g. the position adopted by Descartes – see Chapter 2). In the centre of the continuum is the moral orthodoxy, which summarizes the current moral status of other animals in, for example, the UK. The principle of the moral orthodoxy is that other animals have an interest in not suffering, but this can be overridden for what is deemed to be the greater good of humans (Garner, 2005a, p. 15). At the other end of the continuum we find challenges to both the granting of no moral status to other animals and to the moral orthodoxy, because at this end other animals are granted moral equality with humans. Three divergent positions make up this pole of the continuum. They are the utilitarian position, which argues that other animals should be treated in the same way as humans; the animal rights position, which proposes that other animals like humans have rights, including the right not to suffer and the right to life; and the contractarian position, which argues that other animals have inherent value (Garner, 2005a, p. 18). The positions are explored in more detail in Chapter 8, what I want to focus on here is the continuum of recognition itself.

Knowing the position adopted, not least by sociology and by sociologists, on this continuum is central to our discussion in this chapter as the position taken will inform how that particular sociological stance views the abuse of other animals. The first position is that other animals have no moral worth; so the treatment of them does not register on the scale of morality (Garner, 2005a, p. 15). If we adopted this position we would not consider actions that induce suffering and death in other animals to be criminal acts. We have already seen this in Descartes' work (Chapter 2), where we saw that his presumptions about the lack of sentiency in other animals led him to conclude that this lack 'absolves [humans] from the suspicion of crime when they eat or kill animals' (Descartes, 2007, p. 62). The central position (moral orthodoxy) views other animals as having some moral worth, but not equivalent to that of humans. In discussing the difference in moral worth, Garner differentiates between moral standing and moral status (2005a, p. 15). If we took this position on human moral responsibility to other animals, we would disagree with Descartes because we would view other animals as having moral standing (and thus would consider their interests) but we would see other animals as having less moral status than humans. We can use an example to highlight the position. As discussed in the labelling theory example (above), much of what is done to other animals in biomedical laboratories would be illegal if it took place outside of the laboratory setting (thus, the other animals have moral standing). Because such experiments are often viewed as necessary for human well-being, they are not designated criminal acts when they take place under certain conditions (thus, the other animals have less moral status); but if they take place outside of those conditions then criminal action could ensue (we discuss this in considerable depth when we come to examine experiments on other animals in Chapter 8).

This brings me to the other limit of Garner's continuum. If we took one of the three stances that lie at this end of the continuum, we would argue that other animals should be accorded equal moral status with humans. The utilitarian approach is most obviously associated with the work of Peter Singer (e.g. see Singer, 2002). Singer argues that humans should treat other animals as they would treat other humans, and by this he does not mean that we should treat humans as we would treat other animals, rather he is arguing that 'all things being equal' we should treat other animals as we would humans. 'The Great Ape Project' is a good example of Singer's thinking on this. Singer makes the case that great apes (by whom he means humans, chimpanzees, gorillas and orang-utans) comprise a moral community and thus, thinking about crime, they cannot be killed except in self-defence (2002, p. 128). Consequently to kill a great ape for any other reason than self-defence would be an act of murder. The animals rights approach, for example associated with Tom Regan, argues that other animals have moral rights like those of humans 'and treating them as if they do not possess such rights is a form of prejudice and is unjust' (Kalof and Fitzgerald, 2007, p. 23). The most fundamental of the universal moral rights that apply equally to all holders, or 'moral agents' to use Regan's (1983) terminology, is the right to respectful treatment. Moral agents are bearers of these rights and all bearers of these rights have inherent value, as moral agents have desires, perceptions, memory, and other attributes that indicate that they are 'subject-of-a-life'. Because moral agents have equal inherent value, all are equally valuable and thus a 'criminal is no less inherently valuable than a saint' (Regan, 1983, p. 237) and another animal is not less inherently valuable than a human. Moral agents live in moral communities and these moral communities also consist in those to whom moral agents owe duties, that is other moral agents and moral patients such as babies and other animals who are not morally accountable. Thus experiments on other animals, for example, violate the individual rights of other animals and thus are criminal acts. Mark Rowlands' (2002) work provides a good example of the third stance at this pole of the continuum, that is the contractarian standpoint. In summary, following John Rawls, Rowlands argues that the moral rules for a given society are those that would be chosen behind an imaginary 'veil of ignorance' (2002, p. 57). Those who make decisions about the moral rules should place themselves in a position of ignorance about who they are and about their personal qualities (e.g. whether they are human or other animal, female or male, old or young, criminal or victim) and without knowing the positions they will occupy in the real world (Rowlands, 2002, p. 57). If in the imaginary position of not knowing who we are and what position we hold, we would not want to advocate a legal framework that could lead to our suffering or death (or indeed make such treatment lawful), then it is morally wrong to advocate such a legal framework in the real world.

These positions are especially important when thinking about the abuse of other animals. As we have seen, sociologists such as Durkheim consider crime to be an important moral organizing principle, and acceptance, or not, of the abuse of other animals is an important aspect of the ethical principles that govern society. Much discussion about the abuse of other animals centres on the links between such abuse and the abuse of humans, and this has become an important area of sociological research. Crucially, much less attention has been given to the other animal as the one who is wronged; I centre on this area of sociological research in the next section.

The other animal as a 'wronged subject'

It can be argued that a positive consequence of the current position of moral orthodoxy is that other animals are no longer held morally accountable for their behaviour and thus the prosecution of other animals is a thing of the past (Ritvo, 1987). However, Ritvo (1987) cautions, the consequence is that in their suffering and abuse they are not usually deemed to be victims of crimes. Hence, Derrida claims, 'An animal can be made to suffer, but we would never say, in a sense considered proper, that it is a wronged subject, the victim of a crime, of a murder, of a rape or a theft, of a perjury' (Derrida, 1992 in Beirne, 1999, p. 128). This position is contested (e.g. Singer's case about Great Apes), but it is commonly established. However, if we accept that 'crime' does not have an ontological existence, the role of decision-making in the labelling of some actions as crimes (or as lawful behaviour) and some actors as criminals (or as law-abiding individuals) becomes apparent. Often the context in which an action takes place leads it to be labelled as illegal or as legal. For example, many of the deaths and painful practices meted out by humans to other animals violate anticruelty laws (Beirne, 1999, p. 128). Annually, over 100 million nonhuman animals worldwide are used in experiments (Rowlands, 2002, p. 124) and such experiments cover a range of painful and life threatening procedures involving scalding, breaking of bones and blinding, and many of these maltreatments are inflicted without anaesthesia. Although most of us would define such actions as cruel, they are perfectly legal as they are permitted by law. Similarly billions of fish, chickens, cows, sheep, pigs and other animals are killed every year. Again this practice is entirely legal if it takes places within, for example, the confines of the food production industry. Millions of fish, birds, deer and other animals are hunted each year for sport or for other purposes. This is wholly legal if it takes place within permitted hunts or places for hunting. So literally billions of other animals are legally abused every year and, indeed, Benton argues, 'The largest-scale and most systematically organized abuses of nonhuman animals occur in intensive rearing regimes in agriculture and in research laboratories' (1998, p. 171). So, although cruelty to some other animals is made illegal in the UK by, for example, the Animal Welfare Act 2006 (enacted 2007), cruelty to and abuse of animals is sanctioned by other laws and in other ways. Thus, Beirne argues, 'far from it being a heuristic device for the study of animal abuse, criminal law is a major structural and historical mechanism in the consolidation of institutionalized animal abuse' (1999, p. 129). As criminal law is an inadequate basis for studying cruelty to other animals and as much other animal abuse takes place within the law, Beirne asks, 'how should we understand animal abuse?' (1999, p. 129). Indeed, what do we mean by 'abuse'?

We have already encountered Descartes' notion that other animals are morally equivalent to machines because, he believed, other animals cannot think or communicate. Jeremy Bentham argued against this idea by commenting thus: 'The question is not, Can they reason? Can they talk? But Can they suffer?' (Bentham, 1970 (orig 1789), p. 283). He is clear that at least some other animals feel pain and, as a result, like humans they have an interest in avoiding pain. In consequence, as a utilitarian, Bentham insists that the same consideration should be given to other animals as we give to humans and thus calls on humans to avoid inflicting suffering on other animals. Singer (1990) agrees and takes the position that tackling cruelty and abuse of other animals is part of the important process of making our lives cruelty-free.

Although criminal law often consolidates institutionalized animal abuse (Beirne, 1999, p. 129), legislative efforts have also centred on trying to tackle cruelty to other animals. The US Animal Welfare Act 1966, the UK Animal Welfare Act 2006, and the Animal Health and Welfare (Scotland) Act 2006 are obvious examples, and other countries (such as India) are drafting such legislation. The details contained in these Acts differ from country to country, but they are similar in that they give protection (not rights) to at least some other animals. However, studies of legislative success and failure reveal that the laws are often ineffective or are not taken seriously. For instance, Clifton P. Flynn (2000) observes that violence towards other animals, even in the instances in which it is outlawed, is often overlooked because societies value humans more than other animals. This is compounded by the broad range of socially accepted forms of violence against other animals, which contributes to the lack of interest in forms of violence towards other animals that are socially less accepted (Flynn, 2000, p. 87). Where abuse is recognized, the conviction rates can sometimes be as little as 50 per cent (Flynn, 2000, p. 87). Flynn concludes that 'animals, along with non-verbal human infants, are the only victims of systematic discrimination and exploitation who truly cannot speak on their own behalf' and thus, he argues, adult humans must speak for these exploited creatures (2000, p. 87). Even though abusive and cruel acts against other animals are worthy of discussion in their own right, and violence against other animals is often connected to violence against humans (Flynn, 2000, p. 87), sociology has tended to neglect the abuse of other animals as an acceptable area of academic research. This is because other issues are viewed to be more important and are given higher priority by researchers (Flynn, 2000, p. 87). But this is not confined to sociology, even in criminology there has been 'no explicitly theorized category of animal abuse' (Beirne, 1999, p. 119). Because, as we have seen, criminology is a 'discourse that investigates the whys, the hows and the whens of the generation and control of the many aspects of social harm – including abuse, exclusion, pain, injury and suffering' (Beirne, 2002, p. 381), it seems surprising that the abuse of other animals has been largely ignored by criminology. Flynn puts forward an explanation. He argues that few social scientists give such study high or indeed any priority because the acceptance of the abuse and exploitation of other animals 'is supported by powerful institutions of religion, science, and government' and 'those who are interested in the welfare of animals are perceived as overly emotional or irrational' (Flynn, 2000, p. 87). Of course, Andrew Linzey admits, the ways in which humans treat other animals draw out our emotions, but this does not mean that the human treatment of other animals cannot be the subject of reasoned analysis (2009, p. 1). In this regard, he points to the important role that theology and philosophy have played in the study of our treatment of other animals. Similarly, Beirne (1999) argues that the human abuse of other animals is a vital area for study in criminology, and Flynn (2000, 2011) comments that sociology can make a significant contribution to the study of cruelty to other animals and its links with abuse of humans. In what remains of this chapter I point to a number of ways in which the abuse of other animals is important, and indeed essential, to sociological analysis.

The abuse of other animals as a signifier of conflict among humans

Research on the abuse of other animals points to the ways in which it can be a sign of conflict, discord and violence among humans. For example, other animals who are used as weapons against humans or who are used as signifiers of criminal identity are frequently subjected to cruelty and abuse. In these ways and in more commonplace ways, other animals are considered to be private property who can be owned by humans and who can be treated as commodities. Such research could benefit from more engagement from sociologists.

Other animals as personal and public property

In his call for a zoological connection in sociology, Bryant (1979) suggests various ways in which sociology could fix its gaze on human relations with other animals and one of these is what he calls 'zoological crime'. Here he argues for a study of the way in which other animals are seen as personal property and as public property (Bryant, 1979). This is important not least because this leads to conflict between humans who see other animals as objects over which there is competition and dispute (Beirne, 1999, p. 121). Other animals are viewed as commodities who are owned by humans and who can be damaged, stolen, held to ransom and otherwise misappropriated by other humans, by non-humans and by other means. Indeed, techniques used to label other animal as possessions, such as tattoos, ear tagging and branding, affect the welfare of other animals (Cazaux, 2007). Beirne (1999) observes that this is the most common way in which other animals enter the domain of crime and the field of criminology. In terms of the law the focus is not on the suffering inflicted on other animals, rather it is on those who are labelled the 'victims' of the crimes, that is those who 'own' these creatures and thus are considered to have been harmed by the abuseof the other animals concerned. There are many examples of this in the UK alone, ranging from farmers who are legally permitted to kill dogs who 'worry' sheep, stolen race horses who are talked about merely in terms of their monetary value, and poaching laws that see other animals as 'game' who are owned by landowners. Beirne (1999, 2009) notes that academic studies that explore the abuse of other animals focus on whether or not the perpetrators should be prosecuted and on the class relations that are fundamental to the 'ownership' of other animals. However, Beirne argues, it is important that academic studies try 'to uncover the sources and forms of power and social inequality and their ill effects' (2002, p. 381), which includes study of the suffering of other animals. Consequently academic research could examine why the deaths of billions of other animals that occur in the food and clothing industry and that result from sport, leisure and recreational pursuits do not violate anticruelty statutes (Beirne, 1999).

Other animals as dangerous to humans

Investigations into connections between the abuse of other animals and interhuman violence have been much more prevalent in public and academic (including sociological) discussions about crime and other animals than concerns about the suffering of the other animals involved. For example, there has been progressive news media interest in the UK about the dangers that specific breeds of dogs are reported to pose to humans. For example, in 1994 Anthony Podberscek published research that traced articles about dog attacks on humans by monitoring five major British daily newspapers (four broadsheets and one tabloid) and their related Sunday editions over a five-year period (1988–92 inclusive). He found that between 1989 and 1991 there was intense news media interest in dog attacks on humans whereas in the years 1988 and 1992 there was little. In 1989 and 1990 reports most often centred on the German Shepherd and the Rottweiler, but the Rottweiler suffered disproportionate negative publicity. In 1991, the American Pit Bull Terrier was most often in the news and in the same year the

British Government introduced the Dangerous Dogs Act, which made pit bulls, along with three other breeds, illegal to 'own' without specific exemption. Podberscek (1994) concluded that the media, public and government response was an overreaction to the generally held ideal that the dog's position in society is as a loyal and faithful companion.

One way of looking at the increased news media interest in dog attacks on humans is to examine the phenomenon through the lens of the sociological discussion of 'moral panics'. Sociological thinking on moral panics can be traced to the work of Stan Cohen (2003 (orig 1972)), who coined the term. In his examination of Mods and Rockers youth cultures in 1960s' Britain, Cohen found that there was disproportionate media concentration on Mods and Rockers as intimidators in seaside towns, even though they were involved in very little vandalism or violence. Cohen defined this as a moral panic, that is where 'A condition, episode, person or group of persons emerges to become defined as a threat to societal values and interests...' (2003 (orig 1972), p. 28). Moral panics are predicated in a disproportionate response to the actual threat of violence and 'folk devils' (i.e. those who are constructed as socially threatening) are created through such moral panics. In the 1960s, Mods and Rockers were folk devils who were constructed as threatening the public; in the 1990s specific breeds of dogs became folk devils. Although it cannot be denied that dogs and others animals have harmed and have killed humans, two issues that are largely absent from the media reports, and from subsequent research, are whether the other animals who are identified as dangerous have suffered at the hand of humans and whether justice is served by killing the other animals involved.

The abuse of other animals and human criminal identity

News media stories about 'dangerous' breeds of dogs often centre on the ways in which these dogs are used as identity markers, that is as so-called 'status' dogs or 'weapon' dogs (Maher and Pierpoint, 2011). Jennifer Maher and Harriet Pierpoint (2011) observe that status dogs are often associated with urban youth gangs in the UK. They found that the majority of the youths in their small-scale study who belonged to a youth gang used dogs for protection and companionship, and utilized them to enhance their own status. Over 20 types of abuse of other animals were described by the participants in the research (Maher and Pierpoint, 2011), and it is well known that the abuse of other animals is a serious aspect of antisocial behaviour (Flynn, 2000, 2011). Flynn notes that this is especially so amongst males because, he observes, 'Like some of the other antisocial behaviours, animal cruelty may be rewarded by peer groups, and for males, linked with developing and proving one's masculinity' (Flynn, 2000, p. 88). This is especially evident in men who seek entertainment and amusement from the reprehensible behaviour of pitching dogs against each other in contest fights. These 'dogmen' use dog fighting to validate their masculinity (Evans, Gauthier and Forsyth, 1998). Although the dogmen deny that cruelty and abuse are involved, cruelty and masculinity can be tightly bound. This is summed up in the following comment from one dogman: 'I don't care how long my dog fights, if he's still able to keep going and chooses to quit, he's not coming home with me. He's a dead dog' (Evans, Gauthier and Forsyth, 1998, p. 830).

In their analysis of dogmen in the USA, Craig J. Forsyth and Rhonda D. Evans (1998) used neutralization theory to examine the ways in which these men tried to counter the stigma and criminal identity that is associated with dog fighting. Neutralization theory centres on the defence mechanisms that individuals use to rationalize their deviant behaviour and to neutralize negative accounts of what they do (Sykes and Matza, 1957). Forsyth and Evans (1998) found that the dogmen they interviewed used a range of such mechanisms; for example, they argued that pit bulls are natural fighters and thus to deny them the right to fight would be cruel; they suggested duplicity in those who criticize dog fights but who attend other similar events such as boxing; they gave details of their loyalty to the long history of dog fighting and to dogmen of long standing; and they claimed to be good people (1998, p. 206). Opponents of dog-fight contests counter these mechanisms with accounts of the terrible injuries and suffering sustained by the dogs (Forsyth and Evans, 1998). Forsyth and Evans conclude that 'the dogmen attempt to account for their conduct by considering their criminality and exemplary behavior on the same balance sheet; evil fades because it is outweighed by goods' (1998, p. 215). Paradoxically, in using such mechanisms, the dogmen were indicating that they have a 'strong bond to the conventional moral order' (Forsyth and Evans, 1998, p. 215) because they wanted to be accepted as noncriminal.

The abuse of other animals as a signifier of interhuman violence

There have been a number of studies that link the abuse of other animals in childhood with subsequent human abuse later in life. For example, criminal psychologists have linked childhood cruelty to other animals with adult aggression towards humans (Wright and Hensley,

2003, p. 71). For this reason, Flynn (2000, 2008, 2011) suggests that the connections between the abuse of other animals and interhuman violence makes the abuse of other animals an important area of study for sociologists.

Flynn's (2000) study of 267 introductory psychology and sociology undergraduates at a university in the USA revealed, in keeping with previous studies (e.g. by Miller and Knutson, 1997), that childhood experience of the abuse of other animals is common. Flynn's questions were designed to reveal whether childhood experiences of the abuse of other animals affected adult life. He found that nearly half (49.1%) of those surveyed had witnessed or perpetrated torture, killing or sexual violation of other animals while they were children (Flynn, 2000, p. 88). More male (two-thirds) respondents than female (four out of ten) respondents had experienced the abuse of other animals and most strikingly over a third of males (34.5%) had inflicted violence compared with less than a tenth (9.3%) of females (Flynn, 2000, p. 88). Cats, dogs, rodents, birds and reptiles were the most common victims of the abuse and the most common methods of cruelty were shooting and direct physical aggression like 'hitting, beating, kicking, or throwing an animal against the wall' (Flynn, 2000, p. 89). Retrospective studies confirm the 'progression thesis', or 'graduation hypothesis' (for discussion, see Flynn, 2011), which suggests that children who deliberately harm other animals will 'graduate' to deliberately harm humans, while others confirm the 'desensitization thesis', 'which is more concerned with the general levels of callous behaviour which permeate modern life' (Taylor and Signal, 2008). In a study of retrospective accounts of 64 male sex offenders, David Tingle et al. (1986) found that nearly half (48%) of convicted rapists and 30 per cent of convicted child molesters reported having been involved in childhood incidents of cruelty to other animals (cited in Taylor and Signal (2008)).

Another way in which abuse of other animals is a signifier of interhuman violence is the way in which it acts as a marker of violence in the family (Flynn, 2000, 2008, 2011). For example, Nik Taylor and Tania Signal (2008) show that the abuse of other animals is an indicator of child abuse within the family and Frank R. Ascione's (1998) study reveals that the abuse of other animals is associated with domestic violence. In his survey of 38 women survivors of domestic violence, Ascione (1998) found that the majority of those who lived with other animals as companions (71%) had experienced their abusive partner threatening or harming the creature. Moreover, in 2007 Ascione reported that concern for companion animals affected the decisions of women about whether or not to leave an abusive partner (cited in Ascione,

2008, p. 475). Catherine Faver and Elizabeth Strand found that nearly half of their sample of 41 women survivors of domestic violence who lived with companion animals at the time of the abuse reported that their partners had threatened or had actually harmed the other animals with whom they lived (2003, p. 1367). They concluded that 'service providers should inquire about battered women's concern for their pets and should include arrangements for animals in safety planning' (Faver and Strand, 2003, p. 1367). Such studies reveal that the abuse of other animals should be considered in domestic violence programmes and subsequent protection (Ascione, Weber and Wood, 1997; Faver and Strand, 2003; Ascione, 2008). The keeping of other animals as 'pets' is widespread in many societies, and humans who live with 'pets' often have a strong emotional attachment to them, particularly during stressful episodes such as divorce (Flynn, 2000, p. 92). However, the position of the other animal is not always a happy one because one of their functions is to serve as 'surrogate enemies' in the household (Veevers, 1985) and it is precisely 'their status as family members' that can make them 'more vulnerable to violence' (Flynn, 2000, p. 92). Indeed, Frank R. Ascione's and Phil Arkow's research revealed that children exposed to domestic violence are more likely to abuse other animals than those who are not exposed to such violence (cited in Ascione, 2008, p. 475).

Such research, though extremely important, shows that 'unlike any other form of violence, the published research on animal abuse is motivated almost entirely by its association with violence against people' (Solot, 1997, p. 257). In this regard, Dorian Solot argues that the placing of violence into separate sections that are reserved for, for example, abuse of children, or domestic violence, or cruelty to other animals, obscures the connections between acts of violence and prevents an examination of the abuse of other animals in its own right (1997, p. 257). Different strands of violence are connected in a 'web' of violence (Solot, 1997) and, ultimately, Flynn (2000, 2011) argues, putting an end to the abuse of other animals is an important step in reducing all violence.

Concluding remarks

The abuse of other animals is only labelled as such when it is unauthorized cruelty that occurs outside of sanctioned spheres such as laboratories in which experiments on other animals take place and farms on which other animals are raised for consumption; this is evidence of a system that is loaded against other animals and in which other animals are treated as second-class creatures. Sociology has given relatively little attention to violence against other animals, despite the fact that billions of other animals are subjected to authorized and unauthorized abuse and cruelty every year. Research on the abuse of other animals is often 'anthropocentric and speciesist' because the work that has been done often centres on the links between the abuse of other animals and interhuman violence or, where it does look at harms to other animals, it excludes authorized violence and abuse that takes place in, for example, institutions like farms and laboratories and for purposes of human leisure (Flynn, 2008, p. 170). Such institutionalized violence accounts for the majority of the violence executed against other animals (Beirne, 1999). The 'graduation thesis' is a clear example of the way in which the abuse of other animals is rated as secondary to other forms of abuse, and of the way in which it is only visible as a springboard to actions seen as more worthy of sociological interest. Moreover, where other animals are visible in the legal system they often appear as commodities over which there is human disputation. Such accounts 'betray the predominant fact that they [other animals] are not regarded as sentient beings capable of pain and suffering but as Cartesian automata that are mere appendages to humans' (Beirne, 1999, p. 125). Thus, Piers Beirne and Nigel South (2007) advocate 'green criminology' that, in its focus on the study of crimes against the environment, also includes work on violence against other animals and that views other animals not simply as appendages to humans.

5
Town and Country: Animals, Space and Place

Introduction

In Britain we often picture the countryside as a rural idyll; a landscape inhabited by domesticated other animals (such as sheep and cows) in patchwork fields surrounded by wilder areas populated by red squirrels and deer who live in the woods and forests. In the city or town our encounters with other animals are more likely to be found in walking with dogs, sightings of cats sitting in windows and encounters with welcome visitors (such as blackbirds and robins) or unwelcome creatures (such as gulls and pigeons). If we go out to sea we might hope to come upon other animals such as puffins, gannets, dolphins and seals. Like in all countries across the world the different landscapes and seascapes are associated with a variety of other animals and, depending on where we are, we will have different ideas about the other animals we might meet. Although these ideas are often idyllic, they demonstrate that, as a grouping, other animals are 'subjected to all manner of sociospatial inclusions and exclusions' (Philo, 1995, p. 655). In this chapter I explore spatial relationships between humans and other animals and examine how these are related to power relations. Thus I ask how are other animals included and excluded from different places? In order to answer this question I compare urban areas such as towns and cities with places viewed as natural (which, for ease, I call the countryside), and consider how humans make a distinction between places for other animals and places not for other animals (or not for specific other animals). In order to do this I reflect on the labelling of other animals as 'wild' and 'domesticated', and consider a range of segregated places such as zoos, farms, laboratories and slaughterhouses, and more integrated places such as people's homes. To put this into context I start with a brief discussion about how sociology has examined and conceptualized space and place.

Sociology, space and place

Sociology has traditionally centred on the urban as this is the main space in which humans interact with each other and, consequently, where the subject matter of sociology is seen to take place (Chapter 1). Urban sociology is most usually associated with the structure of the city or town and with the social relationships among humans that take place within these urban areas. Of course, urban space is inhabited by a vast range of other animals, most of whom have been overlooked by sociology precisely because the urban is associated with the human. Although the sociological focus on humans has been seen as a justifiable outcome of the border line between nature and culture, in fact the urban environment cannot be extracted from the natural environment as it is closely related to it. The boundary between nature and culture is not simply conceptualized as a line of separation; Jennifer Wolch argues that notions of improvement are entrenched in conceptualizations of urbanization. As she puts it, 'Urbanization in the West was based historically on a notion of progress rooted in the conquest and exploitation of nature by culture' (Wolch, 1998, p. 119). While urbanization is recognized as a relational process that involves the changing of the natural, the problem with conceptualizations of 'urbanization' in urban theory, she argues, is that this process is regarded as transformative progress that 'develops' 'empty land' into 'improved land' (Wolch, 1998, p. 119). Although other animals were conspicuously absent from the late 1990s' urban theory that Wolch was commenting upon, they were not absent from the land that urban theorists were talking about. Rather than being empty land, urban land is populated by other animals (Wolch, 1998), but these other animals are usually rendered invisible and the negative actions upon them are given little consideration. In consequence, that which is designated 'progressive' in urbanization is constituted by actions taken in the interests of humans alone 'ignoring not only wild or feral animals, but captives such as pets, lab animals, and livestock, who live and die in urban space shared with people' (Wolch, 1998, p. 119). In this way, the 'progressive' or 'civilized' urban space has been counterpoised against the countryside, which has been seen customarily as a degraded and beastly place populated by other animals and degraded humans. For instance, Marx and Engels suggested that in creating cities, the bourgeoisie 'has greatly increased the urban population in comparison with the rural, and has thus rescued a considerable part of the population from the idiocy of rural life' (Marx and Engels, 2009 (orig 1848), p. 9). Her reading of academic accounts of urbanization has led Wolch (1998) to conclude that most scholars of urbanization are anthropocentric in their focus. However, she points to the sociologist Ted Benton as a prominent exception. Benton (1998) is concerned about the effects of urbanization on human connections with other animals. He argues that the growth of towns and cities in the early modern period undermined human familiarity with other animals because significant parts of the urban population became detached not from other animals 'but from direct involvement in the utilitarian and exploitative side of our relationship with them' (Benton, 1998, p. 73). Thus urban humans are often unaware of the realities of the process via which our consumption of other animals takes place (discussed in Chapter 6), and our interactions with living other animals are often confined to walking with dogs, stroking cats, feeding birds and visiting zoos.

Urban space has been so differentiated from the countryside in sociology that the term 'environment' was at one time wholly associated with the rural; in consequence 'the environment' as a concept was seen to be outside of the remit of sociology (Dunlap and Catton, 1979). Since then, conceptualizations of 'the environment' have changed, and the term now encompasses a diverse range of meanings, from the natural world in its entirety to the contexts in which phenomena exist. So we might talk about the 'learning environment', the 'home environment' and the 'shopping environment', as well as the 'urban environment' and the 'natural environment'. However, like in other areas in sociology, the term 'environment' has also retained its association with the 'natural world', and this natural world is now seen as a relevant area of sociological study that often centres on the ways in which its elements are threatened by human technology (e.g. see Beck, 1992; Urry, 2011). This more recent focus has led sociologists to draw important links between sociology and ecological studies that centre on living organisms and the natural environment, and risk continues to be an essential factor of this work also. For example, sociologists have demonstrated how human social activities are causing mounting stress on the 'natural environment' and how these effects are not equally distributed (Marx, 1994). Human activities have also changed our relationships with other animals. In many pre-industrial societies other animals, though eaten, were and are seen as part of the order of things. In contrast, in industrial societies we put the nonhuman world (which includes other animals) at risk because we view it as something to be controlled, shaped and tamed through science and technology. The problems associated with human involvement are viewed to be so intense that there have been suggestions that we are experiencing the 'death of nature'. In this respect Bill McKibben (2003) warns that the changes wrought by humans to every part of the globe mean that nothing is untouched by human hands, and Carolyn Merchant (1982) cautions that advancing technology has resulted in shortages of natural resources. However, the sociological notion of the 'natural environment' as a resource in a realist sense, albeit a 'resource' that is under threat, is one that has been criticized.

Bronislaw Szerszynski, Scott Lash and Brian Wynne criticize the way in which the 'natural environment' has entered sociological thinking because, they suggest, it seems to be seen 'simply as a material substrate of the social, defined by scientific enquiry' (1996, pp. 1–2). Because a realist approach has been applied in such enquiries the natural environment and, we might add, the urban environment have been viewed as having an existence that is independent of our conceptualizations of them. In contrast, a social constructionist approach centres on the ways in which distinctions between the 'natural environment' and the 'urban environment' are socially produced. If we think of the 'natural' and the 'urban' environments as existing in a wider space we can see that 'space is a product of interrelation', that is 'is constituted through interactions' and as such it is 'always under construction' (Massey, 2005, p. 9). So, rather than taking space for granted by seeing the natural environment and the urban environment as areas that exist independent of our conceptualizations of them, Doreen Massey (2005) encourages us to think about the social dynamics that produce space, the way in which space is always in the process of being made, and the implications that space-in-process has for living beings. However, Szerszynski et al. caution against conceiving of space as simply a social construction as they suggest, 'the occasional sociological tendency to criticize such scientific reification by advancing the alternative view that all such problems are "mere" social constructions, and hence (it is implied) not real, is equally misleading' (1996, p. 3). In this regard Adrian Franklin recommends that social constructionists 'should not obliterate the value of conceptualising nature...as an objective reality, a real materiality that exists prior to any social constructions that people may put on it' (2002, p. 25). So, it seems space is real, and how we conceptualize space has real effects. This is crucial because, as Henri Lefebvre makes clear, '(Social) space is a (social) product [...], the space thus produced also serves as a tool of thought and of action [...] in addition to being a means of production it is also a means of control, and hence of domination, of power' (1991, p. 26). This important observation informs the discussion in this chapter. I conceive of space as a phenomenon that is in process and think about the very real effects of this process on other animals and on relations between humans and other animals. Thus I use terms like 'nature' and 'culture', and the 'natural environment' and the 'urban environment' as if they refer to distinct phenomena. This is not to suggest that I accept these distinctions as natural. Sociology understands space as being comprised of relational arrangements of living beings and social goods (Low, 2006); the terminology used assists the examination of the processes involved in categorizing space.

Nature/culture boundary

The chapters up to now have centred a good deal on sociological (and at times, other) ideas about differences between nature and culture. Such ideas include taken-for-granted assumptions about differences between humans and other animals. Humans are said to have culture (which is seen as the defining and differentiating element of being human) and other animals are seen as cultureless natural beings. This nature/culture distinction is not only elemental to notions of the biological foundations of the 'human–animal' binary; but also spatially present in our notions of the difference between the cultured or social environment and the noncultured or natural environment. We often understand the natural and cultural domains to be different and distinct, and these distinctions are central to sociological notions of the spaces that are understood to be human places and those that are considered to be nonhuman places. Even though sociology has traditionally focused on culture, now there is growing sociological attention paid to the relationship between nature and culture (Newton, 2007, p. 1), though still other animals are somewhat invisible in this newly found interest.

Early sociologists who examined human relations with other animals largely saw other animals as an important aspect of 'traditional' societies and often examined human relations with other animals in terms of religion and totemism (Jerolmack, 2007). For example, Durkheim (1965 (orig 1915)) was interested in the ways in which moral authority and social order in traditional societies became embodied in totemic representations of other animals (we explore this further in Chapter 7). However, Bruno Latour (1993) points out that in modern societies social order is viewed as being clearly separated from nature, with the consequence that examining other animals in society is no longer seen as sociologically important (Jerolmack, 2007, p. 75). Latour argues that this creates the fantasy of 'two entirely distinct ontological zones: that of human beings on the one hand; that of nonhumans on the other' (1993, p. 11). Jerolmack (2008) sees this as pointing to at least one important aspect of Latour's analysis. The social construction of nature lies in the fact that our relations with other animals are unimaginable because of the nature/culture dualism that we construct. Latour (2004) contends that examining the ubiquitous relations between humans and other animals would result in us seeing the world as it really is (Jerolmack, 2008, p. 3). Therefore, in order to see society, Latour (1993), like Donna Haraway (1991), argues that social scientists should study society as it is lived; that is as a phenomenon that is comprised of humans, other animals, objects and technologies. For instance a study of the 'natural environment' should recognize that the human 'colonization of nature' has only been possible through the use of technology, machinery and experts such as scientists and administrators (Szerszynski, Lash and Wynne, 1996, p. 3). Despite the increased awareness of such interactions, only a limited increase in sociological interest in human relations with other animals or in interactions between other animals has ensued. This is problematic because humans and other animals are hardly ever separated in spaces deemed to be 'natural' or 'cultural', not least because humans and other animals compete for natural resources in each of these spaces (Buller and Morris, 2007, p. 471). This antagonism is central to a number of interactions that are referred to by Xavier de Planhol (2004) in his catalogue of human interactions with other animals (cited in Buller and Morris, 2007). Planhol's (2004) schema covers a variety of interactions including destructive interactions through which other animals are exterminated or made extinct by humans (e.g. killing badgers for reasons of 'land management'); shared interactions in which other animals come close to humans (e.g. seeing field mice in wheat fields and pigeons in cities); and domesticated interactions (as seen in human relations with farm animals and other animals who are kept as companions) (Buller and Morris, 2007). Despite the range and ubiquity of human interactions with other animals, we are 'so tied up with our visions of progress and the good life that we have been unable to (even try to) fully see them' (Wolch and Emel, 1998, p. xi). However, when we do see them we notice that our relations with other animals are often destructive to them.

Although some interactions between humans and other animals might be seen as positive, destructive interactions are the most prominent and these are mostly associated with urbanization. Jennifer Wolch and Jody Emel (1998) observe that urbanization has disastrous ecological effects, threatening ecosystems and a range of species of other animals and plants. However, this is not to say that pre-urbanized societies worked in harmony with the environment. Small-scale farming, the herding of other animals, and the use of horse drawn ploughs had far-reaching effects on other animals and on the natural environment (e.g. see Thomas, 1983). Even so, high-income urbanized industrial nations place the greatest demand on ecosystems and other animals, and Wolch and Emel suggest the plight of other animals 'has never been more serious than it is today' because:

> Each year by the billions, animals are killed in factory farms; poisoned by toxic pollutants and waste; driven from their homes by logging, mining, agriculture, and urbanization; dissected, re-engineered, and used as spare body parts; and kept in captivity and servitude to be discarded as their utility to people has wanted. This reality is mostly obscured by the progressive elimination of animals from everyday human experience....
>
> (1998, p. xi)

Paradoxically, one of the significant features of post-industrial societies is an increasing concern with the human impact on the natural environment and on the other animals who live there. Nevertheless, developments in human technological power are accompanied by enormous risks to all living things (Beck, 1992), but the human enthusiasm for technological progress has resulted in a loss of sight of its regressive effects on other animals. This can be explained by the nature/culture dualism, which renders human relations with other animals invisible (Latour, 2004) and which is based on the human power to exclude. Foucault's ideas are useful in this regard as he points out that space is never innocent because knowledge and power are linked to the ways in which space is used 'to isolate and exclude, to segregate and thus manage social difference' (Wolch and Emel, 1998, p. xiv). For example, specific other animals are excluded from particular types of places such as urban areas, though such excluded other animals are often accepted by humans who live in the borderlands of 'urban-wildlands' (Philo, 1995). So, it is not only the nature/culture distinction that is an issue, but also the spaces related to conceptualizations of that division. In the remainder of this chapter I explore how other animals are included, excluded and rendered visible/invisible in such spaces. I examine how humans make distinctions between places *for* other animals and places *not for* other animals by considering conceptualizations of other animals as, for example, 'wild', 'domestic' and 'vermin', and by thinking about places such as zoos, farms and laboratories. I consider why some other animals are welcomed in urban areas (e.g. cats and dogs) while others are not welcomed (e.g. rats and pigeons). This will help expose the ways in which humans construct space, and the ways in which such constructions determine the fate of the other animals who inhabit differentiated spaces (Lawrence, 1994, p. 184). Of course, the social construction of nature is not limited to space, other animals are also socially constructed and in their constructions 'the agency of natural and especially animals [is] denied' (Wolch and Emel, 1998, p. xv). Other animals, their habitats and their behaviours are constructed to fit in with our notions of space and appropriate roles within designated places (Tuan, 1984; Yarwood and Evans, 1998). Of course, space is complex and it cannot be divided into discrete places, not least because there are many borderlands; however, for the purposes of what remains of this chapter I refer to space and its relation to other animals in Chris Philo and Chris Wilbert's terms, that is where '...zones of human settlement (cities) are envisaged as places for companions, zones of agricultural activity (farms) for livestock, and zones of unoccupied lands (country side) for "wild animals" ' (2000, pp. 10–11). Consequently, where apposite, I centre on the urban environment, the agricultural countryside and the wild countryside.

Other animals in the urban environment

The boundary between the 'natural' and the 'cultural', which is tied with the inclusion and exclusion of other animals from specific spaces, is especially noticeable when thinking about urban spaces because 'animals as a social group are included in or excluded from the city' (Philo, 1995, p. 655). Cities and towns are not built to accommodate other animals (Wolch, 1998, p. 125) because the urban environment is viewed as a place that belongs to humans (though humans consent to the inclusion of some other animals), while the countryside is seen as a place where most other animals belong. Thus, humans have notions about the proper places that other animals should occupy; however, there is no consensus among humans about how this should play out in practice. For example, Philo and Wilbert (2000) observe that many humans want to be very close to other animals whereas others want to keep other animals as remote as possible. Regarding the latter, they suggest that 'Unless said animals appear in the form of food, the suggestion is that animals should definitely remain "outside" the perimeter of human existence, banished from the vicinity of everyday human life and work' (Philo and Wilbert, 2000, p. 10). Although the focal point of human control of other animals is found in the urban, humans also try to control the countryside, and the control of both these environments extends to the control of national boundaries (where certain other animals are seen as 'foreign' and are pathologized – see discussion below). Consequently our ways of trying to control and organize humans and other animals in space relies on a web of ideas about spatial boundaries associated with, for example, the countryside and the urban, salt and fresh waters, protected and unprotected places, and nationhood. However, the distinction between the countryside and the urban is the most obvious example of boundary work between the human and the other animal. The social construction of the division between urban environments and what are deemed to be the 'natural' environments in the countryside is most apparent when we think about what goes on in each side of the boundary. For example, what we call the natural environment is often made up of farms, zoological theme parks, forests, heritage areas and places designated as areas of natural beauty or as wildlife protection areas; all of which are 'managed' by humans. Thus many so-called natural environments are 'peopled, have histories and geographies, and in being so are in some way or another social as well as natural productions' (Hinchliffe, 2007, p. 11). Similarly, the urban environment clearly cannot be totally managed by humans not least because humans cannot control the skies or the subterranean areas. So, even though the city is not built to accommodate other animals, 'a subaltern "animal town" inevitably emerges with urban growth' (Wolch, 1998, p. 125).

Other animals are intensely affected by urbanization because urbanization, especially within industrialized societies, changes landscapes by denaturalizing them and by polluting air, water and land (Wolch, 1998, p. 125). Some other animals adapt well to the changes (e.g. rats and pigeons), but others have been increasingly driven out of the areas in which they or their ancestors used to live (e.g. deer, badgers and eagles). Urbanization in the contemporary world is based on the exploitation of both the nonhuman (i.e. other animals, plants, land and water) and the human. The problems that urbanization causes for other animals are often so great that humans create, for example, nature reserves (there are 224 national nature reserves in England alone (Natural England,

2011)) and corridors designed to enable other animals to travel safely from one reserve to another (in Southern India a wildlife corridor has been created to enable elephants to travel between two nature reserves in Karnataka (Wildlife Trust of India, 2007)). So other animals inside and outside the urban environment are affected by humans and by urbanization and it is the fusion of the complexity of space with the complex relationships between humans and other animals that maintains hierarchies of power and privilege, not only between humans and other animals but also among humans and among other animals. In order to think about this I return to Mills' (1970) sociological imagination. Arluke and Sanders (1996, p. 168) suggest that if Mills were thinking about human relations with other animals, he would encourage sociologists to look beyond the ways in which humans interact with and treat other animals to the 'general social processes' that lie beyond these. In this regard they point to the 'ladder of worth' and the 'systems of social control' that maintain the hierarchical ranking of humans and other animals (1996, p. 168). This reverberates with the points raised in Chapter 3 as it is those who are positioned at the bottom of the ladder who are most oppressed and exploited. Other animals are on this ladder but they do not all share the same rung. In this regard Arluke and Sanders refer to 'sociozoologic systems' that 'rank them according to how well they seem to "fit in" and play the roles that they are expected to play in society' (1996, p. 169). In this sociozoologic system, 'good' and 'bad' other animals are differentiated; those perceived to be 'good' accept their subordinate role and thus are brought into contact with humans whereas those who are viewed as 'bad' are seen as threats and thus may be killed (1996, pp. 169–70). This sociozoologic system is most palpable in the urban environment as other animals who inhabit or visit urban spaces are often seen as wanted (good) or not wanted (bad).

Domestication and the city

'Good' other animals who live in the city are often (but not exclusively) valued for being 'domesticated'. Identifying other animals as 'domestic' or 'domesticated' requires crossing the boundaries between culture and nature (Russell, 2002, p. 285). This crossing of boundaries is not just spatial and social; the nature/culture divide is also located in assumptions about biological differences. For this reason 'domestication', Nerissa Russell (2002) observes, is very difficult to define, but she discusses a few suggestions. A simple definition refers to the control of the movement and of the breeding of domesticated other animals (Russell, 2002, p. 287). More complex is Sandor Bokonyi's (1969, p. 219) suggestion that domestication involves '...the capture and taming by man of animals of a species with particular behavioral characteristics, their removal from their natural living area and breeding community, and their maintenance under controlled breeding conditions for profit' (Russell, 2002, p. 287). Although Bokonyi points to *exploitation*, others, for example Terry O'Connor (1997), characterize the relationship between humans and domesticated other animals as one of *symbiosis*. Thus, rather than viewing domestication as the human exploitation of other animals, O'Connor sees the relationship as one of equal partners that serves to benefit both humans and other animals (Russell, 2002, p. 289). Pierre Ducos rejects ideas about symbiosis as he argues that humans impose domestication on other animals, because domestication requires other animals to be 'integrated as objects into the socioeconomic organization of the human group, in the sense that, while living, those animals are objects for ownership, inheritance, exchange, trade, etc....' (Ducos, 1978, p. 54 in Russell, 2002, p. 290). Although taming is mentioned by a number of commentators, domestication is often differentiated from taming. Taming is usually seen to be founded in a one-to-one relationship between a particular person and a particular other animal 'without long-term effects beyond the lifetime of that animal', whereas, 'domestication is a relationship with a population of animals that often leads to morphological and behavioural changes in that population' (Russell, 2002, p. 286). This demarcation points to a gamut of attempts to differentiate among other animals that encompasses, for example, associated categorizations such as 'tame', 'tameable', 'domestic', domesticated' and 'semi-domesticated'. There is by no means a consensus on what these differences are and indeed on the definitions of each of these distinctions. For the purposes here I use Jean-Pierre Digard's (1990) broad view of domestication that centres on control and domination of other animals (Russell, 2002, p. 291). For Digard, all other animals who are subjected to the 'domesticatory action' of humans (e.g. 'pets' and 'farm animals') can be discussed within the remit of 'domestication'. What is important about Digard's work is that he centres on the process of domesticatory action that is exerted by humans on other animals in specific contexts and spaces, rather than on specified other animals as being in a state of 'domestication' (Russell, 2002, p. 291). For Digard, Russell notes, 'the degree of domestication varies according to the inherent suitability of the animal species and the technological and social features of the human society' (2002, p. 291).

The domesticated other animals who are closest to humans are those who are kept as 'pets'. Here it is useful to refer to problems with terminology. In much sociological and other academic literature 'companion animal' has replaced the term 'pet'. Arluke and Sanders note a problem with this alternative terminology as 'companion animal' implies mutuality in the relationship between the human and the other animal whereas 'pet' implies a relationship that subordinates the other animal (1996, p. 171). In this chapter I use the term 'pet' as I want to capture the subordination referred to by Arluke and Sanders.

'Good animals': 'Pets'

Arluke and Sanders observe that groups of humans appear on every rung of the sociozoologic ladder and 'good animals' are close to those humans who are at the top (1996, p. 170). It might be argued that 'pets' (unlike other 'good animals' who are on lower rungs) enjoy a close association with humans because they are brought into the urban environment for reason of their affective usefulness to humans (Arluke and Sanders, 1996, p. 170). In this regard, Franklin comments that 'the relationship with pets is the closest and most humanized of human–animal relations' (1999, p. 84) and humans often express love and affection for these other animals who they have admitted into their homes. Keith Thomas (1983) looks back to the early modern period in England and notices the increasing profusion of 'pets' in the cities; that is a profusion of other animals who were not eaten by humans, who were given personal names and who had co-residency with humans. He explains that the cities were considered to be the centres of culture and civilization, and movement to these areas was largely confined to the wealthy. Because these wealthy city dwellers had money and were among the first to be spatially separated from other animals, keeping other animals as 'pets' became fashionable among them (Franklin, 1999, p. 13). As a result, a 'pet' became a marker of refinement because in the city 'animals were less likely to be functional necessities and [...] an increasing number of people could afford to support creatures lacking any productive value' (Thomas, 1983, p. 110). 'Pets' are often treated as members of the family, but Lucy Hickrod and Raymond Schmitt (1982) caution that we should not go too far down the road of seeing 'pets' as *equal* family members. Although 'pets' are invited into human family spaces there are what Hickrod and Schmitt call (1982) pervasive 'frame breaks' that call into question the footing of the 'pet' as a member of the family rather than as an 'animal'. Obvious examples are signs in restaurants that say 'no dogs allowed' and signs in parks that tell people to 'keep dogs on leads' (e.g. see Arluke and Sanders, 1996, p. 12).

Even though 'pets' are often regarded as friends or family members, they are always subordinate to humans (Arluke and Sanders, 1996, p. 171). Yi-Fi Tuan (1984) traces the intrinsically unequal power relations that are central to 'keeping pets'. He asks: 'How have humans established their dominance over beasts that in the wild were too large and fierce to be managed?' (1984, p. 107). Dogs are among the most popular other animals who are kept as 'pets' in the West (Anderson, 2003). This is because, Tuan suggests, the dog is the 'pet' *par excellence* as the dog exhibits 'uniquely a set of relationships we wish to explore: dominance and affection, love and abuse, cruelty and kindness' (1984, p. 107). In order to be accepted into the home dogs, like other 'pets', 'must learn to be immobile – to be as unobtrusive as a piece of furniture' (1984, p. 107), and to stay they must continue to be obedient. Tuan suggests that the exercise of power is part of the attraction of bringing dogs and other 'pets' into the home as 'Power over another being is demonstrably firm and perversely delicious when it is exercised for no particular purpose and when submission to it goes against the victim's own strong desires and nature' (1984, p. 107).

National boundaries shape conceptualizations of who is and who is not regarded as a 'pet', for example, the country in which a dog lives guides whether he or she is treated as a 'pet' or as some *thing* else. For example, in Cambodia, China, Thailand and Vietnam dogs are eaten as food, though they are also kept as 'pets', whereas in countries such as the UK and the USA the idea of eating dogs (and cats) is viewed as abhorrent and morally corrupt (Podberscek, 2009, p. 617). Thus spatial differences influence the social construction of other animals, the rung they occupy on the sociozoologic ladder and their destiny. Values and behaviour associated with spatial differences can also lead to antagonisms between humans. For example, in South Korea the eating of dogs is strongly linked to national identity, and calls from countries in the West to ban the custom are interpreted by South Koreans as an attack on their culture (Podberscek, 2009, p. 615). Nevertheless, distinctions in the status and treatment of other animals are not limited to differences among nations. Within national boundaries, on a race track dogs are no longer seen as 'pets', rather they are viewed as 'racing machines' (Arluke and Sanders, 1996, p. 12). This is exemplified in the official rather than anthropomorphized name that is usually given to a racing dog (examples include 'Flash', 'The Rocket' and 'Little Zippity') and a customary lack of display of affection from humans (Arluke and Sanders, 1996,

p. 12). In an airport dogs can act as sniffer dogs, in factories as guard dogs, and they can be viewed as experimental equipment in scientific laboratories. The distinctions are not really about the dogs themselves, they are indications of the ways in which the social construction and treatment of the dogs is spatial. Dogs in these places are treated as tools rather than as 'pets', and other animals who are used as 'tools' come further down the sociozoologic ladder.

'Good animals': Tools

On a lower rung than 'pets' on the sociozoologic ladder are other animals who have a valuable place for humans in the social order as they are treated as tools (Arluke and Sanders, 1996). Unlike 'pets' who might be thought to be happy with their lot, 'good animals' who are treated as tools are not necessarily happy about the way in which they are treated as their position is based purely on their utility to humans (Arluke and sanders, 1996, p. 170). Arluke and Sanders' (1996) discussion about other animals who are used in experiments exemplifies the point, and I will come to these unhappy creatures in due course, but first I want to turn to zoos and aquariums as spaces in which other animals are treated as tools.

Zoos and aquariums

With the exception of domesticated other animals, humans in the West live largely isolated from other animals. However, we still want to see other animals and this is one of the motivations for holding other animals captive in zoos. Consequently such places, John Berger argues (2009), are 'living monuments' to the disappearance of other animals from human life. Zoos are places where 'that which is culturally defined as nature's "wildness" is brought in' (Anderson, 1998a, p. 120) and thus, although trying to emulate wilder places, zoos are not wild places at all. Zoos are places where the natural is 'introduced into the metropolis and converted into domestic spectacle' and, as such, zoos are presented as testaments of human triumph over nature and of progress on wilder places (Anderson, 1998b, p. 34). So zoos (and their watery aquarium counterparts) are places that transform 'nature' into 'culture'. Nevertheless, Randy Malamud suggests, 'zoos are not a microcosm of the natural world but an antithesis to it' (1998, p. 30). He draws on Berger (2009) to make the point that although zoos are places where 'people go to meet animals, to observe them, to see them, [they are], in fact, a monument to the impossibility of such encounters' (Malamud, 1998, p. 30); humans could not meet other animals in this way in wilder spaces.

In the West zoos have changed. Zoos used to consist mainly of cages that held captive 'exotic' other animals from distant countries. In visiting these places 'people were frequently reminded of prisons, the sadness of separation and punishment, and the high value placed on freedom, liberation and openness' (Franklin, 1999, p. 62). Consequently, before changes were made, zoos were anthropocentric in representation and orientation (Franklin, 1999, p. 74). However, Franklin suggests, two innovations have occurred since the 1970s that have transformed zoos. Endangered animal zoos, he argues, are no longer spaces devoted wholly to spectacle, rather their main purpose is to support the repopulation of wilder places though the breeding of endangered or rare species of other animals (Franklin, 1999, p. 74). For Franklin the second innovation is found in zoos where the main purpose is to mimic complete ecosystems in an attempt to 'display not individual animals but communities of species that had evolved together' (1999, p. 77). Such innovations have led to the conviction that zoos are better now than they used to be, not least because in these remodelled places the other animals are no longer simply tools of spectacle rather they are treated as educational tools for humans, and their confinement is viewed as being for their own good. Such assertions are criticized. Malamud is unequivocal; he argues 'zoos have historically demonstrated their unconcern, or incompetence, with regard to achieving any beneficial impact upon the natural world outside their gates and cages' (1998, p. 46). To emphasize his point he asks: 'Can a polar bear be happy in a warm climate? Can a tiger live a natural existence in a cage or a "terrain"? Does an ocelot really enjoy the passing crowds looking at it all day?' (Malamud, 1998, p. 48). The benefits of the educational agenda so often attached to zoos have also been doubted. For example, Lori Marino et al. argue that that there is no convincing evidence that zoos and aquariums promote positive views of other animals and a concomitant visitor interest in the conservation of other animals or of wilder places (Marino, Lilienfeld, Malamud, Nobis and Broglio, 2010, p. 126). In consequence, although the elimination of cages must have improved the lives of some of the other animals in zoos, the other animals who live in zoos are still held in captivity. Even if we were to accept that such captivity is for the greater good of their cousins who live in wilder spaces, research suggests that the endangerment and the lack of educational benefits put in serious doubt the use of such other animals as tools.

But zoos are not the only places where species of 'wild' other animals are used as tools in the urban environment; their counterparts or once 'wild' relatives are also brought into laboratories. However, unlike zoos, laboratories are places in which captive other animals cannot be viewed by the public; the lived-reality of their utility to humans is thus hidden from most humans. As such, the place that is the scientific laboratory gives us an impression 'of a whole hidden geography' (Philo and Wilbert, 2000, p. 2).

Laboratories

Other animals in laboratories are conceptualized as tools for students and researchers. Often the same species of other animal can be thought of as a tool in a laboratory context while being viewed quite differently outside of that context. For example, Michael Lynch (1988) observes that 'laboratory animals' can be constructed as 'wildlife' in the natural environment or as 'pets' if they are in the home. However, in the 'laboratory animals are progressively transformed from holistic "naturalistic" creatures into "analytic" objects of technical investigation' (Lynch, 1988, p. 266). In consequence, this 'analytic animal' has become nothing more than a tool and he or she is other than the 'naturalistic animal' who is thought of by lay people (Lynch, 1988, p. 266). In this way, Arluke and Sanders (1996) observe, the naturalistic animal is stripped away from the 'lab' animal even before he or she enters the laboratory. This is because huge corporations, such as Du Pont, breed and market other animals, such as mice 'patented' as OncoMouse™ as 'superobjects' who meet the needs of scientists and other researchers. As such these other animals are very much seen as 'good animals' but not in the same way as 'pets'; in order to be 'good' they are treated very badly indeed. Rats and mice who are bred for use in the laboratory have been subject to decades of selective breeding (Birke, Arluke and Michael, 2007, p. 22). In 2010 in the UK alone over two and a quarter million mice and over three hundred thousand rats were used in experiments (Home Office, 2011, p. 20). Unlike dogs and cats, who are also used in laboratories in the UK (5695 dogs and 187 cats in 2010 (Home Office, 2011, p. 20)), rodents, rats and mice are less likely to be viewed as 'pets' in other spatial locations (though increasing numbers of rats are kept as pets in the UK). For this reason rats and mice are usually considered to be more acceptable 'tools' in the laboratory, though in order for them to be so used they have been transformed from 'wild' into domesticated other animals who are 'denizens of the lab' (Birke, Arluke and Michael, 2007, p. 21). This transformation from wild to domesticated has led to a transformation in perceptions among scientists from 'perceiving animals as similar to those in the wild, to seeing them as tools of the trade' (Birke, Arluke and Michael, 2007, p. 21). Thus, the spatial change is preceded by, or results in other animals who look very different from our conceptualizations of them (e.g. a mouse with a human ear), who are genetically different (e.g. genetically modified other animals who are predisposed to human diseases) and whose treatment is very different from how we expect living other animals to be treated (e.g. in having their organs mutilated, their bones broken, their skin burned, their eyes contaminated by pollutants and their bodies corrupted by a range of poisons and diseases). In this way other animals who are naturalistically regarded as 'wildlife' or who are loved as 'pets' are viewed in the laboratory as sacrifices who are transformed into data (Lynch, 1988; Birke, Arluke and Michael, 2007). I discuss experiments on other animals in more detail in Chapter 8. The purpose of drawing them into the discussion in this chapter is that they show that the ontological status of other animals is superseded by the social construction of them and these social constructions are spatially dependent and have consequences for them. The rat in a laboratory is a suffering device who is viewed as a 'good animal' because she or he acts as a valuable tool for humans; however, the freer and better off rat in the city is conceptualized as a 'bad animal', as 'vermin'.

'Bad animals': Pests, contagion and pollution

Despite unrelenting exterminations, urban areas host 'a shadow population' of other animals (Wolch, 1998, p. 119). These other animals are often viewed as 'bad animals'. They are feared by humans and are often constructed as repulsive creatures. For these reasons 'society may ignore, marginalize, segregate, or destroy them' (Arluke and Sanders, 1996, p. 175). In stating that 'bad animals' are those who are seen as 'freaks that confuse their place, vermin that stray from the place, or demons that reject their place' (1996, p. 175), Arluke and Sanders elucidate the association between place and the classification of other animals. This association is also clear in Colin Jerolmack's analysis as he argues that 'the construction of animals as problems relies upon cultural understandings of natural culture relationships, which in turn entail "imaginative geographies" ' (2008, p. 72). As we have already seen, modernity sets a firm boundary between nature and culture, and Jerolmack notes the designation of appropriate places for other animals within this division, in which they are 'experienced as "out of place" and often problematic when they are perceived to transgress spaces designated for human habitation' (2008, p. 72). Other animals who are seen as transgressors often come to be seen as 'vermin' or as 'demons'.

Vermin

Other animals who humans see as vermin rarely physically threaten humans, yet they invoke feelings of disgust as they are often seen as symbols of dirt (Arluke and Sanders, 1996, p. 178). In the 1960s the anthropologist Mary Douglas (1970) demonstrated the process by which a phenomenon comes to be thought of as 'dirt'. She argued that dirt does not have an ontological status rather it is an *idea*, because the label 'dirt' is reserved for matter that is out of place. This observation can be applied to conceptualizations of other animals as vermin. For example, rats are viewed as clean and hygienic 'good animals' when they are purpose bred for the laboratory; they are viewed as unclean and filthy 'bad animals' when they roam free in urban streets and gardens. This conceptualization of rats as vermin also explains the relative acceptability of rats (e.g. in contrast to other animals thought of as 'pets') for use in laboratories (Birke, Arluke and Michael, 2007, p. 25), and this paradox confirms that space affects our conceptualizations of rats. The *verminization* of rats has aided the verminization of another much maligned creature in the urban environment, the pigeon.

Although pigeons have been domesticated by humans for many years, at present they are most likely to be deemed 'out of place' if they inhabit spaces close to humans in cities and towns (Philo and Wilbert, 2000). Pigeons who inhabit urban streets cannot be justly described as 'wild' or as 'domesticated' because they are descendents of domesticated pigeons who escaped captivity and who have adjusted so that their 'natural' state is now one of coexistence with humans (Jerolmack, 2007, p. 75). Pigeons are much maligned. For example, Trafalgar Square in London was once famous for the pigeons who flocked there, and tourists flocked to feed these tourist 'attractions'. In 2000 the Mayor of London banned the feeding of pigeons in the square. Once seen as bringers of peace they are now seen as bringers of disease and contamination. Jerolmack (2008) explores the process by which pigeons have been so problematized. He contends that 'pigeons have come to represent the antithesis of the ideal metropolis, which is orderly and sanitized, with nature subdued and compartmentalized' (Jerolmack, 2008, p. 72). The pigeon has been conceptualized as a health problem though: 'the pigeon's primary "offense" is that it "pollutes" habitats dedicated for human use' (Jerolmack, 2008, p. 72). The anxieties associated with this bird are condensed in the then London Major's description of pigeons as 'rats with wings'. This metaphor, argues Jerolmack, 'reflects a framing of pigeons by claims-makers that renders them out of place in the cityscape' (2008, p. 72).

Moreover, this metaphor relies in an acceptance of the notion that rats are worthy of their maligned status. Rats are unquestionably the most universally maligned species of other animal 'as they are viewed as carriers of diseases and as damagers of crops and food stores' (Lovegrove, 2007, p. 217). So, in the freedom of the city the 'bad' rat is seen as quite different from their unhappy cousins, the 'good' rats in the laboratory, and any other creature who is called a 'rat' is a bad creature who is so maligned.

Demons

Arluke and Sanders argue that lower than vermin, and the lowest on the sociozoologic scale, are demons, other animals who are viewed as the worst of all (1996, p. 180). Unlike vermin, demons are seen as dangerous fiends who might kill humans. Pit bull terriers are a demonized group of other animals. Arluke and Sanders suggest that 'To a large segment of the public, pit bulls are more than dangerous killers; they are demons that eat humans' (1996, p. 184). Pit bulls are one of four breeds of dogs who are listed as dangerous under the UK Dangerous Dogs Act (1991) (Chapter 4) (the others are the Japanese Tosa, the Dogo Argentino and the Fila Braziliero). As with the other three breeds, it is illegal for a person to keep a pit bull in the UK unless he or she is registered on the Index of Exempted Dogs and the conditions of exemption are strictly adhered to. Although pit bulls originated in the UK (the three other breeds have their origins in other countries; spatial issues abound) they are seen as out of place in the UK because they are viewed as being too dangerous. This demonization transcends national boundaries as pit bulls are restricted breeds in the USA and in many other countries. Hence, pit bulls seem to be so demonized that they are seen as not fit to live at all. They are deemed unfit to live in urban environments, where other 'pet' dogs live; they are deemed unfit to live in the countryside, a space felt to be more appropriate for other animals; and they are deemed to be unfit to cross national boundaries as they are illegal in many countries of the world. Pit bulls are on the lowest rung of all; like all demons they are lower than vermin and they are the most feared of other animals.

Other animals in the countryside

Human interaction with other animals in the countryside has had disastrous effects on other animals. Nevertheless, although urbanization progresses with little regard for them (Wolch, 1998, p. 119) concerns about other animals are increasing, evidenced by, for example, fears about the growing extinctions of species through habitat loss. Ironically, such concerns have led to increasingly diverse human involvement in the countryside (by which I mean rural agricultural areas and wilder places). In consequence the countryside, once seen as a natural place, is now seen as a place of scientific fieldwork, conservation and capitalist exploitation (Philo and Wilbert, 2000, p. 1). The countryside is viewed as the space in which most other animals belong and sociozoologic scale of Arluke and Sanders (1996) is an effective way of examining the human constructions of other animals who live in this space. Like in urban spaces, some other animals in the countryside are constructed as 'good' whereas others are viewed as 'bad', and the human species of animal has an unresolved place on the scale; humans must be tired of climbing up and down the ladder that they have constructed. For example, if we think about habitat loss, climate change and resulting extinctions of other animals it is quite clear that because humans are held to blame they are 'bad' animals in this regard. However, human scientists and conservationists who are trying to reverse this trend, or at least to quell the damage that is being done, might be on one of the 'good' animal rungs. Consideration of the place of humans on the sociozoologic scale points to our values about other animals. Other animals who are trying to live in the wilder places but who are endangered or threatened in some way are those whom humans are trying to save and, because humans are trying to save them, they occupy a high rung on the ladder.

'Good animals': Rare, endangered and protected species

Because humans have recast, enclosed and changed wilder places, humans have become increasingly fascinated by them. This fascination is not only associated with fear, control and domination, but also connected with concern, affection and interest (Tuan, 1984). This is evidenced by human concerns about the plight of rare, endangered and threatened species of other animals. The range and number of other animal species on the endangered list is truly staggering and any internet search will show thousands of sites devoted to the plight and fight of endangered other animals. The World Wildlife Fund 2010 annual list of the most threatened species includes those they describe as some of the most 'well-known and beloved species' on the planet (World Wildlife Fund, 2009). Most are endangered because of human exploitation of the other animals concerned and of the spaces they once occupied. For example, 'poaching', deforestation and reduced habitats have led to the tiger and the Javan rhinoceros being endangered. Blue fin tuna are at risk of extinction because they are widely used as human food. Reductions in habitats mean there are only around 720 mountain gorillas left in the world. Deforestation has threatened the survival of the monarch butterfly. There are fewer than 2500 giant pandas living in the wild. Despite these figures, Charles Bergman has reservations about human attempts to quell or reverse such endangerments because, he suggests, 'Too often, the fact that an animal is endangered or threatened constitutes the explanation of why that creature is important. We like animals because they are endangered, as if being endangered confers prestige and status' (1990, p. 4). Bergman is also concerned about those other animals who are not seen as rare or endangered and thus are outside of this remit of 'importance'. The process of the designations of 'importance' and of 'good' and 'bad' come into sharp focus when looking at the position of the squirrel in the UK.

Conceptualizations of the rightful place of the red squirrel and of the grey squirrel encapsulate, in a nutshell, human constructions of valued and not valued 'wild' other animals. Although many squirrels live in urban environments, red squirrels, in particular, are associated with wilder places and, in distinction to grey squirrels, they are seen as threatened and as important creatures. Grey squirrels are conceptualized as the bad 'other' compared to red squirrels; an excluded marked category that is defined as lower to the category that is not so marked (see Chapter 3). Hilda Kean observes that humans who see the countryside as in need of their management have 'depicted the grey squirrel as a foreign force inimical to a mythical English way of life as epitomized by the red squirrel' (2001, p. 163). In England, red squirrels live freely in the wild in a restricted number of places, such as on the Isle of Wight and in parts of Cumbria (Kean, 2001, p. 163). However, in the past they lived in abundance across England and humans shot them as 'pests'. Their disappearance in the wild has become an issue of increasing concern, so much so that the 'red squirrel has been consolidated as a motif of England's heritage' (Kean, 2001, p. 164). Kean observes that 'The red squirrels, despite their previously acknowledged faults of destruction of trees, were constructed as an established symbol of an idyllic rural Britain' (2001, p. 167). As an image of national identity the red squirrel has become constructed as an indigenous creature who is threatened by the 'foreign menace' of the grey squirrel (Kean, 2001, p. 164). So, it is not the creature with the shotgun who has decimated their population, it is the creature with the bushy tail who is being blamed. Referencing the much maligned and most hated of other animals, grey squirrels have been referred to as American tree rats (Kean, 2001, p. 165). The reference to spatial differences is obvious. The grey squirrel is out of place because he or she is over here. Thus, Kean concludes, 'the explanation for the popularity or vilification (or protection) of certain animals at different times owes less to the behaviour of particular animals and more to broader political, social, and cultural concerns in human society' (2001, p. 166) and these relate to notions of belonging.

'Good animals': Domestication and the countryside

In countries like the UK domesticated other animals used for the production of food and other consumer products (such as 'meat' and wool) are usually completely removed from the urban environment (Franklin, 1999, p. 41). This process of separation was particularly marked in the first half of the nineteenth century, and by the early twentieth century 'the entire process of production, slaughter, butchery and packing was an entirely rural activity' (Franklin, 1999, p. 41). In consequence, large areas of land once seen as wild were converted to land associated with agricultural production (Wolch, 1998, p. 124). Domesticated other animals are central to current forms of agricultural production but, Timothy Ingold (1988) notes, domesticated other animals are often regarded as inferior to their wild counterparts, perhaps lacking the souls possessed by wild animals. Thus, they are often seen as having lost their nature because their contact with humans has denaturalized them (Ingold, 1988). Nevertheless, though on a lower rung on the sociozoologic ladder than endangered or protected 'wild' species, domesticated other animals are certainly viewed as 'good animals' not least because domesticated other animals are potent symbols of rural places and associated ways of life, and endangered breeds have become powerful symbols of cultural heritage (Emel, Wilbert and Wolch, 2002, p. 409).

Although we view domesticated animals as part of the rural idyll, their slaughter is not something that is witnessed by many humans. The slaughter of other animals continues to be seen as especially noxious and thus as too repulsive for the urban environment. The removal of the slaughterhouse from the urban environment reflects the desire to make invisible the reality of slaughter (Franklin, 1999, p. 41). This invisibility serves to maintain our view of agricultural rural space as being comprised of places where 'calves nuzzle their mothers in a shady meadow, pigs loaf in the mudhole, and chickens scratch in the barnyard' (Mason and Finelli, 2006, p. 104). This might be our mental picture, but Jim Mason and Mary Finelli (2006) make clear that the reality is very different because the majority of farms are industrial type farms.

I discuss industrial farming in more detail in the next chapter. What is important here is that, like the other animals who are treated as tools in the urban environment, domesticated other animals in the rural environment are 'good animals' because, in their role as being largely instrumental to humans, they are valuable (Arluke and Sanders, 1996, p. 170). Although 'wild' other animals are viewed as superior, if they threaten domesticated other animals they are usually viewed as 'bad animals'.

'Bad animals': Hunting and killing

The countryside is usually viewed as the space for other animals and many of the other animals who reside there are viewed as 'good animals'. In order to preserve these 'good animals', hunting and killing of others labelled as 'bad' is advocated as a method of conservation. Hunters claim to be 'motivated by a passion for ecology, a love of nature (and animals) and a need to conserve animal habitats...' (Franklin, 1999, p. 119). Their argument is that without hunting specific environments could not support the numbers of other animals from particular species, with the consequence that the 'environment will then suffer through overgrazing, and the animals will die of starvation and disease' (Rowlands, 2002, p. 167). This conservation role is supplementary to another common defence of hunting; that is that hunting is essential for keeping down the number of 'pests' (Rowlands, 2002, p. 164) or 'bad animals' in the wild. This is a common argument put forward for fox hunting in Britain, which is also couched in terms of conservation. Nevertheless, Rowlands argues, the motivation is not conservation, rather foxes are viewed as 'pests' because their 'activities adversely affect the interests of farmers' (2002, p. 164). But this ambivalence about hunting motivations demonstrates that the fox has an ambiguous relation with humans because foxes are not just seen as 'pests'. Marvin notes that for centuries in Britain the fox has been chiefly represented as vermin who kills chickens and other domesticated animals (2002, p. 143), but this is when the fox transgresses human space thresholds and enters agricultural land. In other wilder places foxes are viewed as skilled and astute creatures. In these wilder spaces the fox is 'good' and is welcome to eat other wild animals 'but if it "chooses" not to restrict itself to what are perceived to be the "proper" sources of food for a wild creature' (Marvin, 2002, p. 144) the fox becomes 'bad'. So, hunters contend that hunting is more than 'pest' control, and this has another dimension. In hunting the fox, Garry Marvin suggests, hunters are attempting to strike up and develop a relationship with 'a free wild animal whose role is to evade being the focus of attention and, if drawn into a relationship, to attempt to disengage' (2001, p. 278). In reality, human hunters do not directly engage with foxes at all, rather they enact engagement through the foxhound who has been developed for this purpose (Marvin, 2001). Banned under the Hunting Act 2004, fox hunts in Britain have historically depended on the use of foxhounds. This dependence, Marvin contends, is cultivated in the human desire to see 'a contest between a wild animal and a culturally created one' (Marvin, 2001, p. 278), a 'contest' that is played out in spatial relations between 'good' and 'bad' other animals.

Concluding remarks

Conceptualizations of space and allocations to 'appropriate' places are fundamental to power relations between humans and other animals. In this chapter I have sought to consider the ways in which other animals inhabit spaces in human societies and the ways in which they are excluded as 'other' from specific places. This is a particular issue in urban spaces, though it is not restricted to them. Specific groupings of other animals are welcomed and included in the city whereas others are excluded. These exclusions and inclusions are fundamental to the exploitation of other animals. Thinking about our relations with other animals in terms of the sociozoologic scale reveals that those other animals who are labelled 'good' are so specified because of their utility to us, and this is often a function of how valuable they are to us. 'Bad' other animals are those who transgress their designated space. This exposes the ways in which space is fundamental to the designations we give to other animals. 'Pests' are merely other animals who we see as out of place, 'tools' are simply other animals who are moved into spaces in which they can be utilized for human purposes. Other animals are 'minded actors' (Arluke and Sandeers, 1996, p. 42) who cannot simply be regarded as victims of humans as they have territories that they have adapted in some way (Philo and Wilbert, 2000, p. 13). The fox is one such other animal who is recognized for his or her prowess, however if the fox transgresses their designated space she or he could become victimized as a 'pest'. It seems that humans miss their meetings with other animals and thus they hunt other animals, or they capture other animals for display in zoos and aquariums, or they go to 'family farms' and animal sanctuaries to encounter them, or they go into the countryside to see them in their 'natural' habitats, or they try to save them by participating in conservation schemes or in ecotourism. Yet, humans relate with other animals in a range of other hidden spaces including laboratories, industrial farms and slaughterhouses. Other animals kept as 'pets' are the most likely to be invited into urban spaces; some of the most vilified are those seen as 'vermin' who are killed with products specifically designed for extermination. All of these encounters are components of human consumption, which is the subject matter of the next chapter.

6
Consumption of the Animal

Introduction

A major way in which humans experience other animals in their everyday lives is through the consumption of food. Many humans eat 'meat'-based products, drink milk and consume related products such as cheese and butter. Not only are parts, pieces and derivatives of other animals present in many of the foods that humans eat, they are ubiquitous (while often invisible) components of a range of other consumer products such as wool in fabrics and furnishings; fur in clothing; feathers in pillows and cushions; skin in shoes, coats and bags; bone in glue, ornaments and jewellery; and gelatine in photographs and capsule casings. Shopping is the second most popular leisure activity in Britain and other animals are a major component of what we buy. From the living 'pet' for the home to the dead creature on the plate, our shopping experiences are replete with other animals, and increasing wealth along with the concomitant growth of consumption has led to an increased demand for other animals for consumption.

Sociology examines consumption as a phenomenon by centring on the growth of consumer societies. Consumer societies are based in gross materialism (Ritzer, 1993), an approach to consuming that centres on the satisfying of our needs rather than on the ethics associated with what and how we consume (Bauman, 1993). This chapter explores the ways in which the consumption of other animals is central to the development and expansion of consumer societies. Essential to this development and expansion is the augmentation of human choice. In this chapter I investigate increasing choice as it is associated with other animals as products (e.g. the ubiquity and affordability of an increasing range of 'meat'-based products) and the growth of 'ethical shopping' (e.g. associated with vegetarianism and veganism and buying organic foods). There are countless areas of consumption that I could have chosen to look at; however, in this chapter I concentrate on the use of other animals for food. Of course, consumption is directly linked to production as without production there would be little to consume so, as I go through the chapter, I explore how the consumption of other animals is central to the production of consumer products. The chapter begins with a brief overview of how sociology has explored the broad area of consumption.

Consumption, sociology and other animals

Sociology used to be much more concerned with production than consumption; hence conflicts, discriminations and issues associated with paid work were a major focus. However, as consumption has come to characterize many societies (Featherstone, 1991), consumption has become a central feature of sociology. 'Consumption is a cultural universal' (Smart, 2010, p. 4) and it defines Western societies in particular (Ritzer, 1993). In the West a prominent feature of consumption is shopping, which as a pursuit in itself has become a major leisure past time. Shopping is not restricted to functions of, for example, locating food or finding clothes, because places to shop (such as malls) are places to be seen and what we buy defines who we are; as a result, shopping is a major part of identity. Thorstein Veblen's (1994) concept of 'conspicuous consumption' is instructive in this regard as the term refers to the ways in which goods and services are no longer consumed for their utility; rather they are consumed as a demonstration of social status. For instance, wearing the fur of specific other animals, such as minks, has been associated with the rich and the consumption of 'meat' has been a symbol of relative wealth. Such identity markers are also found in the dominance of branded products. Naomi Klein (2001) observes that we live in a branded world where the brands we buy define who we are. Someone who wants a hamburger might have a 'Bigmac'; a traveller waiting in an airport lounge does not simply carry a 'leather' bag, they carry a 'Gucci' or a 'Prada'; a person's coat is not just a fur coat it is 'Gaultier' or 'Armani'. In addition, the high street has given way to the website, which has brought a range of once-inaccessible commodities within our reach. Consumers can buy a range of, at times, illegal living other animals and products derived from dead other animals, which are not available on the high street. In terms of the illegality of this trade, the Convention on International Trade in Endangered Species (CITES) currently protects over 33,000 species of other animals and plants (PAW Scotland, 2011), but still there are many breaches of the law.

Other animals are a ubiquitous yet often invisible element of many products (e.g. wine clarified by fish and pharmaceuticals and other products that have been tested on other animals). In addition, products such as 'meat' are plainly hewn chunks of other animals or are composites of pieces of other animals, yet humans can be 'emotionally reluctant to recognize the embodied nature of meat foods' (Franklin, 1999, p. 126). So, although humans have become spatially detached from other animals (Chapter 5), this is actually in terms of interactions with *living* other animals. Humans are spatially attached to dead animals, in the products and pieces of other animals that they consume. Every time someone pulls on 'leather' shoes, fills their 'leather' case, buttons their 'leather' jacket, sits in a 'leather' seat or eats a 'bacon' sandwich they are engaging with another animal. Indeed, often products made from other animals are considered to be of the highest quality. Thus, a person might judge the 'genuine leather' product to be better than the UPVC version, the merino wool scarf or cashmere jumper to be preferable to the acrylic counterpart and the touch silk to be better than the pull of polyester. In doing so they might not make the connection between the products and the other animals from whom they came, but, even if they do, they may think that in their manufacture (such as products made from wool or drinks made from milk) other animals are not harmed. Although our consumption of other animals extends far beyond 'meat' and dairy-based products, 'Nowhere is the changed relation between humans and animal in modernity clearer than in transformations in the agricultural industries' (Franklin, 1999, p. 126).

Other animals, products and language

Up to now, whenever I have used the term 'meat' I have located it within quotation marks. Sociology encourages us to challenge that which is taken-for-granted; the term 'meat' is highly problematic, yet the word is usually used without critical thought. The role of language in power relations has been closely examined in sociology (e.g. Van Dijk, 1997). Language contributes to the oppression and exploitation of other animals and, Arran Stibbe observes, this is evident in the way we talk about consumer products like 'meat' (2001, p. 145). In Stibbe's words, 'It is in the manufacturing of consent within the human population for the oppression and exploitation of the animal population that language plays a role' (Stibbe, 2001, p. 147). Singer describes the way that the words used reflect the prejudices of those who use them (1990, p. vi). Regarding the term 'meat' for example, he notes how the term distances us from animal suffering because 'The very words we use conceal its [meat's] origin, we eat beef, not bull... and pork, not pig...' (Singer, 1990, p. 95). Kenneth Shapiro puts it simply, 'We do not say "please pass the cooked flesh" ' (1995, p. 671). Stibbe notes that this distancing is also obvious in the words used for the killing of other animals. He observes that 'Animals are slaughtered, humans are murdered. Interchanging these two – *You murdered my pet hamster* – is comical. *The refugees were slaughtered* means that they were killed brutally, uncaringly, and immorally' (Stibbe, 2001, p. 150 original emphasis). Adams (1990) refers to this distancing as the 'erasure of animals'. She argues that by talking of 'meat' rather than bloody pieces of pigs, lambs, cows and calves we participate in language that disguises what is happening (1990, p. 67). For Regan, using terms like 'meat' and 'pork' draws our attention away from the other animals concerned and their uniqueness, and contributes to 'the system that allows us to view animals as our resources' (1985, p. 13). Viewing other animals as resources is further compounded by discourses associated with farming, for example in the use of terms like 'livestock', 'fish stocks' and 'animal agriculture'. Such terms, argues Adams, 'eliminate the animals as animals; instead they become bearers of food' (1990, p. 68).

Production and consumption

The flesh of other animals is a central component of the western diet (Franklin, 1999, p. 148). For example, in 2010 the annual consumption for individuals in the UK was, on average, 76.2 kg (excluding fish), up from 69.9 kg the year before (The Vegetarian Society, 2010). There are global differences. People in developing countries currently eat, on average, one-third the flesh products and one-quarter of the milk products compared to the developed countries in the richer northern hemisphere (Delgado, 2003, p. 3907). This difference is a sign of relative wealth rather than of food predilections (Franklin, 1999, p. 147). However, this is changing, as we will see below. Within areas of flesh consumption there are differences: Europeans prefer to eat cows; Australians prefer to eat sheep; and though humans in these areas of the world eat fish, fish and shellfish are more central to Japanese diets (Franklin, 1999, p. 147). However, the consumption of flesh other than that of fish and shellfish is rising, because of the growth of Japanese international tourism and the fashion for what is seen as 'exotic' food (Franklin,

1999, p. 147). Global production of 'meat' alone is projected to double from 229 million tonnes in 1999/2001 to 465 million tonnes by 2050, and milk production is set to rise from 580 to 1030 million tonnes in the same period (Wilkie, 2010, p. 10). Rhoda Wilkie argues that the increasing demand for such other animal-based products is the result of the three factors: growing global population, countries becoming richer and an increased preference for animal-based products (2010, pp. 10–11).

In order to cope with growing demands there have been historical changes in farming, with a move from 'animal husbandry to animal industry' (Wilkie, 2010, p. 7). This is most notable in the Fordist approach, which is designed to raise the affordability, the profitably and the consumption of other animal-based products (Franklin, 1999, p. 130). This approach is applied to a range of such products, for example 'meat', eggs, milk and milk-based produce. The main tenets of the Fordist approach centre in intensifying production while at the same time minimizing labour costs. Maximizing profits means that there is little interest in the welfare of other animals or for those in the workforce; thus, Stibbe (2001) argues, those who work in the industry are encouraged to focus on profits and to neglect the suffering of other animals. Accordingly the welfare demands of care are seen as 'little more than a commercial cost' (Stibbe, 2001, p. 154). However, this conceals a complexity of relationships; for example, Wilkie found that people who worked in the field of the reproduction of other animals tended to express varying degrees of emotional attachment to and concern for the other animals whilst workers who were involved in the killing of other animals expressed varying degrees of emotional detachment (2005, p. 13). Nevertheless, the welfare of other animals is usually, at best, a secondary consideration though, Garner points out, the terrible realities of industrial farming are often hidden by proclamations of concern for the welfare of other animals (1998, p. 463).

Many of the most horrifying intensive farming practices have made media headlines (Wilkie, 2010); the cruelty is obvious and animal welfare groups and organizations have headed up campaigns against such practices. For example, the incarceration of calves in veal crates in which they could barely move made headline news in the UK. These crates were banned in the UK in 1990 and in the EU in 2007 and some states in the USA are phasing in such bans (Compassion in World Farming, 2008). Nevertheless, the changing conditions that intensive farming brings for the other animals concerned are evidenced in more routine everyday foods, for example in the changes in milk production in the USA. Dairy farming in the USA is more industrialized today than ever before. Don Blaney (2002, pp. 2, Table 1) notes that in the USA in 2000 a total of 167,658 million pounds of milk was produced, up from 116,602 million in 1950. Yet, the number of cows used for milk diminished over the same period, down from nearly 22 million in 1950 to just over 9 million in 2000. On average each cow produced 5314 pounds of milk in 1950 compared with 18,204 in 2000 (Blaney, 2002, pp. 2, Table 1). In order to produce this amount of milk cows must give birth every 12 or 14 months and each cow must produce around 14 litres of milk a day. Dairy cows on such farms are exhausted and many experience terrible health problems associated with incessant birthing and milking. This is just one example; Franklin cites a number of further shocking examples of intensive farming practices. Battery egg production is founded in the keeping of five hens in cages 45×50 cm in size which are stacked in tiers of between three and five cages. The labour costs are much reduced as, compared with free range farms that need one worker for every 1000 chickens, only one 'stockman' is needed for 30,000 chickens (Franklin, 1999, p. 136). Chickens described as 'broilers' are only six or seven weeks old when killed, yet they weigh the same as chickens who are double the age. In order to encourage such weight increase the chickens are kept in 24-hour lighting, and arsenic and growth hormones are added to their food to encourage them to eat more and more (Franklin, 1999, pp. 137–8). Piglets are weaned at the very young age of three weeks, and within five or six days of being weaned their mother is subjected to forced insemination so that she can give birth to the next set of offspring. For the duration of her pregnancy she is kept in a two-feet-wide windowless stall with full light (Franklin, 1999, p. 141). No consideration is given to the pain, anguish and suffering caused by these repeated pregnancies (Stibbe, 2001, p. 34). As for the piglets, when four months old they are said 'to produce the best pork' whereas 'bacon is usually from six month old pigs' (Franklin, 1999, p. 141). Franklin observes that only sheep and goats have yet to be subjected to intensive farming practices (1999, p. 147). More recent technological changes, for example associated with genetic modification, mean that selective and intensive farming practices can be taken further still (Twine, 2010).

In order to produce the amount of animal flesh and animal products that is wanted for consumption, Stibbe explains that human populations must implicitly consent to the mass incarceration of other animals in intensive farms and the industrialized killing that follows this confinement (2001, p. 145). This is revealed by the willingness of humans to buy other animal products sourced from such places. In this way the association between production and consumption is clear. The production of products made from other animals is inextricably grounded in the consumption of other animals. Thus, the differences in sociology between consumption (i.e. the process by which goods and services are used to satisfy economic needs) and production (i.e. the creation of things) are blurred for other animals as other animals are consumed by producers in the production of other animal products and they are then consumed by consumers who buy other animal products. So, the 'coercive power used to oppress animals depends completely on a consenting majority of the human population who, every time it buys animal products, explicitly or implicitly agrees to the way animals are treated' (Stibbe, 2001, p. 147). This is especially visible when consent has been withdrawn, as is evident in boycotts of veal, battery farm eggs, and all flesh products by vegetarians and all other animal-based products by vegan consumers (Stibbe, 2001, p. 147). However, when consent is given, consumers of other animals often seem unaware of the farming practices they are condoning and of the other animals whom they are eating, or if they are aware they are 'emotionally reluctant to recognize the embodied nature of meat foods' (Franklin, 1999, p. 126).

'Meat' in the western diet

Although the consumption of flesh is central to the western diet, Franklin (1999) cites changes that have affected its position. Transformations have occurred in the culture of eating 'meat', in overall levels of consumption of 'meat' and in the presentation of other animal-based foods. Moreover, there are frequent reports about the association between other animal-based foods and health risks (Franklin, 1999, pp. 147–8), and in the West there has been a growth of vegetarianism. I will look at each of these in turn.

'Meat' production

Historically 'meat' eating has been associated with wealth and status (Franklin, 1999, p. 148). However, the shift to intensive farming associated with Fordism has meant that 'meat' has become both cheaper and more readily available. The high plains in the USA, known as the 'beef belt', have farms with 112,000 cows (Wilkie, 2010, p. 10). Australia is the world's seventh largest 'beef'-producing country and it is the world's second largest exporter of 'beef'; 67 per cent of total 'beef' and 'veal' production is exported from Australia (RSPCA Australia, 2008). Such intensive farms produce more 'meat' than European farmers can imagine and the scale is so great that the flesh of other animals has become 'a daily food for most people, as opposed to the occasional "special" meal' (Franklin, 1999, p. 148). This is not to say that there is no intensive farming in European countries like the UK. The British Hen Welfare Trust (2010) estimates that there are 16 million battery hens currently in cages in the UK. There are also moves towards the mega-style farming of other animals. For example, at the time of writing *Midland Pig Producers* is seeking permission to build the UK's largest ever pig farm containing 25,000 pigs, including 2500 sows 'producing 1000 pigs a week for sale' (Midland Pig Producers Limited, 2011). The deleterious effects of such industrial farming practices on other animals are considerable. Regan notes that 'The vast majority of these animals, literally billions of them, suffer every waking minute they are alive. Physically, they are sick, plagued by chronic, debilitating diseases. Psychologically, they are ill, weighed down by the cumulative effects of disorientation and depression' (2004, pp. 89–90). Garner points out that 'the general public in many countries would be prosecuted for keeping animals in some of the ways they are commonly kept by farmers', which illustrates the taken-for-granted approach to the suffering of other animals (2005b, pp. 106–7). The ways in which other animals are treated before they reach the plate means that diet and cuisine have also become global, environmental, political and ethical issues (Franklin, 1999, p. 148). For example, there has been a substantial rise in the popularity of organic animal-based products. Currently an estimated 3.2 million households in Britain buy some organic 'meat', and in Wales alone organically managed agricultural land increased annually at a rate of more than 20 per cent per year between 2006 and 2008 (Frost, Smith and Francis, 2010, p. 3). Organic farming prides itself in the establishment of a more ethical relationship between farmers and consumers, other animals, land and the production of food, and as such it aims to provide other animals with 'natural' living conditions, where they have space to move around, to exercise and to graze, and to eat naturally (antibiotics and other drugs are prohibited) (Holloway, 2003, p. 147). Other animals who live on such farms are considered to be better off as they are healthier and more contented. In consequence, organic animal-based products are viewed to be better for other animals and, because the other animals are healthier, better for the humans who consume them. Consumers have responded. In 2008 in the UK alone the value of the retail organic 'beef' market was £41.7 million and the retail organic 'lamb' market was £18.5 million (Frost, Smith and Francis, 2010, p. 5).

'Meat' consumption

The recent economic recession led to a contraction of markets for 'meat'. For example, in 2009 the world trade in cows and sheep for flesh for human consumption saw a downturn (Frost, Smith and Francis, 2010, p. 3). However, reports now reveal upward movement as demands for 'meat' in China and India rise (United States Department of Agriculture, 2011). So, although the actual consumption of 'meat' in richer nations is higher than that in poorer nations, the growth in 'meat' consumption in relatively newly developing countries has outstripped that in the richer nations threefold (Delgado, 2003, p. 3907). One reason for this is increased affluence. The richer nations experienced an expansion in the consumption of 'meat' in the 1970s (Franklin, 1999). As incomes in poorer nations rise above poverty levels and as the populations become urbanized they tend to consume more 'meat' (Delgado, 2003, p. 3907). But it is not only affluence that has led to changes. The introduction of intensive farming offers increased supply and lower prices which has led to increased demand. In addition to being a status food that is associated with wealth, 'meat' had for some time been a favourite food that people wanted to eat daily. In addition, up to the 1970s it was assumed that 'meat' was essential for health (Franklin, 1999, p. 151) and milk had a similar reputation. There was little attention given to the risks of eating 'meat', 'a theme that was going to dominate the meat industry in the 1980s' (Franklin, 1999, p. 151), but since that time such risks have resonated across other industries associated with the production of other animal-based foods. The staple breakfast food, the egg, became associated with high cholesterol levels and research showed that 'Cardiovascular disease grew directly as a result of the new meat-rich diets' (Franklin, 1999, p. 151), especially diets rich in red 'meat'. Red 'meat' has also been linked with a range of cancers including bowel and stomach cancer. The most significant reason for the decline in red 'meat' eating among men is its connection with coronary heart disease (Franklin, 1999, p. 159). Thus, in the 1980s 'meat' eating decreased for all classes and the highest income groups reduced their rates of consumption more markedly than other income groups (Franklin, 1999, p. 152). So, in the richer nations, it seems that the consumption of 'meat' has been transformed from being a high-status healthy practice to a practice associated with poor health and lower income groups. However, the changes conceal some trends in increased 'meat' consumption. For example, the consumption of newer 'exotic meats' such as emu and kangaroo has increased (Franklin, 1999, p. 152) as has the consumption of 'meats' associated with healthy lifestyles, such as fish.

Until the 1980s health fears about the risks from 'meat' and 'meat' products were largely confined to heart disease and cancer; however, since then a range of media reports have led to fears about infectious diseases that are said to have entered the food chain through the intensive farming of other animals for the production of their flesh for food (Franklin, 1999, p. 162). A case in point was the link made between the human brain disease sporadic Creutzfeldt-Jakob disease (CJD) and bovine spongiform encephalopathy (BSE) found in cows whose flesh was sold for human consumption. Consequently, Franklin observes, modern 'meat' eating, particularly since 1945, 'has been shaped by the Fordization of every process in the animal industry, from birth to burger' and food scares have had the effect of putting the other animal back in 'meat' as consumers come to see that the processes used to 'make meat' were in fact putting them at risk (1999, p. 162). Franklin concludes, 'These rationalised, intensive processes, so studiously hidden from the public gaze, were revealed to be the source of the new risk. In short, the new methods of meat production rendered all meat a potential health risk and it lost its innocence as a marker of modern progress' (Franklin, 1999, p. 164).

Such scares have also been seen elsewhere, for instance the links between salmonella in eggs and salmonella-infected battery chickens. However, 'meat' eating in particular has been laden with possible dangers to human health. Consumers have increasingly been spatially removed from the living other animals they eat (Chapter 5). In consequence, 'Modern consumers are highly dependent on large numbers of unseen others in the agriculture and food industries and their willingness to consume these products hinges on trust' (Franklin, 1999, p. 165). Rather than being associated with health, food is now commonly viewed as a 'pathogen, a source of disease and ill-health' (Lupton, 1996, p. 77) and this is especially true of 'meat'. Being spatially removed from the living creatures they eat as dead 'meat', consumers are unable to ensure the healthiness of the food they consume and hence the state has taken over the role of giving such assurances (Franklin, 1999, p. 173). This has a long history. In light of concerns from UK consumers about BSE, 'meat' containing spinal nerve materials (e.g. T Bone 'steak') is now banned in the UK (Franklin, 1999, p. 172). Of course, 'To the consumer, risk is not actually about probabilities at all. It's about the trustworthiness of the institutions which are telling us what risk is. Do we believe them?' (Jacobs, 1996 quoted in Tester, 1999, p. 216). Many do not. As a result some consumers have become vegetarians for health reasons. Because of its 'experiential naturalness', vegetarianism and veganism are ways of consuming food without depending on experts who are seen as unreliable (Tester, 1999, p. 216). This is because vegetarianism and veganism are perceived to be healthy diets due to their association with natural methods of production (Tester, 1999, p. 216). However, although free from 'meat' or free from other animals in total, these dietary options have not been free from health scares. For example, in 2011 a couple in France was prosecuted because the death of their baby was linked, concluded the deputy state prosecutor, to the deficiencies that the mother's vegan diet had imposed (Willsher, 2011). Matthew Cole and Karen Morgan point out that such 'vegaphobic' discourses in the news media, which associate vegans with outlandishness, faddishness and the killing of dependent others through 'neglectful' dietary ethics, serve to hide the violence towards and killing of so many other animals for human food (2011, p. 149). Accordingly the vegan ethical stance, so ridiculed by the media, is transformed into a risk to human health. In contrast, vegetarianism is often associated with dietary health and thus for many becomes a lifestyle rather than an ethical choice. This contention will be discussed below.

The presentation of 'meat'

Although 'meat' is a daily food for many humans, and food scares have increased public awareness about the origins of 'meat', the ways in which it is packaged and marketed means that 'meat' is often not associated with the other animal who was killed for its production. Ground 'beef' is a form of 'meat' that is extensively consumed in the West, but it is a form that distances the ground flesh that is consumed from the bodies of cows from whom it comes. Americans eat on average more than 14 pounds of ground 'beef' yearly, in foods such as burgers and pasta sauce (Davis and Lin, 2005). Many of the consumers of such foods are children, yet research shows that children often do not know the bodily origins of the 'meat' they consume (Franklin, 1999, p. 148). The way in which 'meat' is produced has been increasingly removed from the public gaze. Huge intensive industrial farms are not part of our vision of the rural idyll and nor are they what we generally see when we go into the countryside. Moreover, as we saw in Chapter 5, slaughterhouses have been removed from sight. Indeed, notes Franklin, the euphemism 'abattoir' serves to distance us symbolically from the killing that takes place in the slaughterhouse (1999, p. 156). Noélie Vialles' (1994) study of slaughterhouses in south west France led her to conclude that a complex system of avoidances promotes a denial of what goes on in them and this system, which centres on a disassociation from death, determines the way in which humans prepare the bodies of other animals for human consumption. 'Meat' is a particular case in point because, in turning the body of another animal into food, the transformation renders the dead body socially acceptable via a range of processes (Franklin, 1999, p. 153). Because consumers do not want to see the whole dead body, butcher shops have changed from displaying a whole 'dressed carcass' to the display of cuts of 'meat' that cannot be identified with the whole other animal (Franklin, 1999, p. 148). 'Meat' is often packaged in small portions in supermarket freezer cabinets, which serves to 'distance the consumer from all connections with the animal of origin and reduce meat to a more abstract notion of animal protein' (Franklin, 1999, p. 155). This is followed by the preparation and the cooking of 'meat', which turns the flesh of the other animal from a bloody lump of muscle into a tolerable mouthful. In the process of cooking, Claude Levi-Strauss (1964) observes that which is natural undergoes a process of socialization and thus becomes more acceptable to humans for food (Franklin, 1999, p. 153).

The growth of vegetarianism

There has been a growth of vegetarianism, especially among young educated consumers (Franklin, 1999, p. 148). Taking the UK as an example, although the vast majority of people in the UK eat 'meat', over the last ten years the number of vegetarians has more than doubled (The Vegetarian Society, 2010). Most recent statistics show that there are now 1.8 million vegetarians in the UK, accounting for 3 per cent of the population (The Vegetarian Society, 2010). Research points to four reasons for becoming vegetarian: personal health worries, concern about cruelty to other animals, concern for world hunger and a green agenda that is associated with broader environmental concerns (McDonald, 2000, pp. 2–3). Health scares associated with 'meat' and the consumption of other animal-based food products (discussed above) are a considerable motivator for giving up 'meat' and/or dairy products (MacNair, 2001). However, ethical concerns remain a major factor as 'compassion for animals is one of many motivations for becoming vegetarian or vegan' (MacNair, 2001, p. 64). Franklin distinguishes between the motivations of those who reduce 'meat' consumption and the motivations of those who become vegetarian. Referring to research undertaken by N. J. Richardson et al. (1994), Franklin points out that concerns about the welfare of other animals account for up to four-fifths of the reasons for vegetarianism, whereas the most commonly given reason for reducing 'meat' consumption is concern about own health (1999, p. 162). In addition, the desire to lose weight or to stay slender is a major factor in reduction of 'meat' consumption, especially among women (Franklin, 1999, p. 161). Nevertheless, it seems that ethical considerations continue to be especially central to the adoption of a vegetarian (and vegan) diet. This is evidenced in the 'McLibel' trial in the UK and the global movement against the fast food company McDonald's that ensued. George Ritzer and Elizabeth Malone (2000) give an account and an analysis. In 1990 the fast food chain McDonald's sued five activists in Britain for libel, in what turned out to be the longest trial in British history. McDonald's sued the activists over their alleged distribution of a leaflet titled 'What's Wrong with McDonald's: Everything They Don't Want You to Know'. The leaflet accused McDonald's of cruelty to other animals, of destruction of the environment, of exploitation of employees and of selling unhealthy food. The website campaign against McDonald's (and other such companies), called *McSpotlight*, reported an average of 1.75 million 'hits' a month (Ritzer, 1993, p. 113). At the time of writing this website reports that McDonalds is 'the world's largest user of beef and thus is responsible for the slaughter of hundreds of thousands of cows per year. In Europe alone they use half a million chickens every week' (McSpotlight, N/D). McDonald's is an immensely powerful and influential global enterprise. It has such a commanding presence that Ritzer (1993) has used its tag in his analysis of the process of domination and change in contemporary society, that is 'McDonalization'.

'McDonaldization' and vegetarianism

McDonald's bases it approach in 'efficiency, predictability, calculability and control' and in doing so it epitomizes the Fordist and automated approach to production (Ritzer, 1993). When entering a McDonald's restaurant the consumer knows they will get their food fast (it is efficient), it will conform to what they expect it to be (it is predictable), the 'Bigmac' will consist of two 1.6 oz 'beef' burgers (it is calculated) and the preparation and packaging of the food will be done with machine-like precision (it is controlled). Although 'The McDonald's fast food factory is represented as the end, in effect the fulfilment, of this particular series of rationalizing processes' (Smart, 1999, p. 4), Ritzer (1993) refers to the 'Irrationality of Rationality' as it relates to McDonald's. Ritzer and Malone define 'McDonaldization' as the process by which the principles of efficiency, predictability, calculability and control associated with McDonald's 'are coming to dominate more and more sectors of American society and an increasing number of other societies throughout the world' (2000, p. 99). Although Ritzer suggests that the principles upon which McDonald's is based are an exemplar of the contemporary development of rationalization (e.g. the unsentimental and calculative thinking associated with the efficiency of technology), he argues that McDonaldization has also created irrationalities that have led to inefficiency, unpredictability, incalculability and loss of control. For example, McDonald's is accused of causing environmental hazards and suffering to other animals on intensive farms; is charged with dehumanizing customers and staff; is criticized for 'false friendliness' because the outlets and the fastness of the food limit or eradicate genuine friendliness between staff and customers; and is blamed for causing disenchantment because the same products are offered wherever one goes (Ritzer, 1993). McSpotlight is not the only resistor to this process. The contributors to Barry Smart's edited volume *Resisting McDonaldization* (1999) discuss a range of oppositions to the effects of McDonaldization. In one of the chapters Keith Tester (1999) centres on resistance through vegetarianism.

Tester (1999) argues that Ritzer ignores the moral questions associated with McDonaldization. He puts his challenge as follows: 'I want to propose that Ritzer's account of McDonalization mirrors the moral emptiness which rational organizations require and promote: Ritzer's account of McDonalization is itself as morally empty as the world it seeks to describe' (1999, p. 208). In order to explore what he sees as this moral emptiness Tester poses the question, does vegetarianism resist McDonalization and rationalization? (1999, p. 208). In noting that there is no essential link between rationality and morality, Tester cites Zygmunt Bauman's observation that 'all people involved in the work of an organization follow the commands they receive and are guided only by them... and that means that people should not be diverted by their personal beliefs, by their convictions and by their emotions' (Bauman, 1994 in Tester, 1999, p. 208). From this, Tester concludes, moral values have become, perhaps, 'a problem to be overcome in the processes of rationality' (1999, p. 208). Thus, it seems that moral values are less and less (and perhaps are not even) considered with the result that victims are overlooked and concerns about them are discarded.

Tester centres on the moral values associated with the welfare of other animals, and in order to explore such values he grounds his discussion in Singer's contention that 'If a being suffers there can be no moral justification for refusing to take that suffering into consideration' (1976, p. 9).

For Singer, if such consideration is given to humans it is speciesist to withdraw other animals from such consideration. Thus, Singer argues that the recognition of speciesism requires us to confirm individually to the genuineness of our concern for other animals (1976, p. 175). For Tester, the most sincere and comprehensible demonstration of this concern is vegetarianism (1999, p. 212). Tester agrees with Singer; vegetarians bring together 'conduct and ethics'. Vegetarians are not merely indulging in a dietary preference, they are conducting their lives around an 'appreciation of what is taken to be a moral fact that the human interest in eating meat is less important than the preference for an animal to live a life without the experience of unnecessary pain' (Tester, 1999, p. 212). He concurs with Singer's argument that a boycott of 'meat' is essential, in order to avoid collusion with cruel and intensive farming practices and to avoid contribution to the profits made out of such practices (1976, p. 175). Consequently vegetarianism is a form of boycott that 'turns the individual into the self-aware author of her or his own moral integrity and ethical being' (Tester, 1999, p. 213). Despite its ethical provenance vegetarianism, as we have seen, is sometimes adopted for health reasons. Tester contends that if the reasons are grounded in health 'the ethical dimension of vegetarianism is cut away and all that remains is a dietary form which promises to offer some kind of insurance against the risks generated by factory farming and other technological processes' (1999, p. 217). Although the outcome might be the same in that vegetarian consumers eat 'meat'-free foods, for Tester there are two distinct types of vegetarianism: 'ethical vegetarianism' and 'lifestyle vegetarianism' (1999, p. 217). Ethical vegetarianism is that which is promoted by Singer; however, lifestyle vegetarianism, suggests Tester, is the most prevalent form in the West (1999, p. 218). Tester is scathing because lifestyle vegetarianism 'replaces the *being* of the ethical conduct of life with the *doing* of the consumer. It is in this way that lifestyle vegetarianism is easily compatible with the relationships and procedures of a McDonaldized environment' (1999, p. 218). Most contemptible of all, he suggests, 'vegetarianism itself is liable to become McDonaldized' (Tester, 1999, p. 217). How does Tester reach this conclusion? Returning to the Fordist approach adopted by McDonald's (i.e. efficiency, predictability, calculability and control), Tester sees efficiency in the idea that vegetarian food sustains a healthy body, calculability in the emphasis on nutritional content, and predictability and control through standardized recipes and vegetarian convenience food. In sum, life style vegetarianism makes money (Tester, 1999, p. 219). Tester notes that the value of the 'meat'-free market in 1995 was £100 million (1999,

p. 219); figures for 2008 reveal that the market is worth a staggering £739 million (The Vegetarian Society, 2010). One of the biggest brands is Linda McCartney which sells 31.28 million vegetarian sausages every year in the UK alone (Connell, 2011). Tester observes that this is a long chalk from Singer's call for vegetarianism (1999, p. 219). The split in the meaning of vegetarianism (ethical and lifestyle) problematizes any oppositional intent it might have to McDonaldization (Tester, 1999, p. 119). For Tester, vegetarianism only stands as a form of resistance when it is not a question of 'what's in it for me' (Tester, 1999, p. 220). Tester concludes, 'Lifestyle vegetarianism puts us back in the iron cages of the McDonaldized worlds. It makes meat avoidance totally rational. Meanwhile ethical vegetarianism makes us morally rigorous but, for the most part, utterly marginal' (Tester, 1999, p. 220). The problems that Tester has found with vegetarianism are, according to the research, cleared away by veganism. According to Josephine Donovan (1990), veganism is the ethical consumer's response to the suffering of other animals. Gary L. Francione defines veganism as 'much more than a matter of diet, lifestyle, or consumer choice; it is a personal commitment to nonviolence and the abolition of exploitation' (2008, p. 16). He seems to agree with Donovan as, he suggests, '... if we are not vegans we certainly *are* animal exploiters' (Francione, 2008, p. 17). This is not just an academic approach. Cole and Morgan note that participants in research projects cite ethical considerations as the major motivation for their own veganism (2011, p. 135). In consequence, Cole and Morgan suggest, 'it is therefore plausible to assert that on the basis of existing evidence, veganism is understood by most vegans... as an aspect of antispeciesist practice' (2011, p. 135). This is evident in the eco-feminist position posited by Adams, who states: 'I am a vegan-feminist because I am one animal among many, and I don't wish to impose a hierarchy of consumption upon this relationship' (Tyler, 2006, p. 123).

Concluding remarks

Humans consume billions of other animals every year. From the living 'pets' who are loved to the dead cows who are eaten, from the 'leather' in shoes to the living bodies used for product testing; from the fur in a coat to the wool in a scarf; other animals are there for the taking – at a price. The cash price paid by the consumer can amount to a bit of loose change; the price paid by the other animals is usually the foregoing of their contentment and the loss of the lives. Other animals, both living and dead, are treated as commodities. Intensive industrial farming is one of the most brutal manifestations of this commodification, and the restaurant chain McDonald's is an obvious example of the rationalization of the consumption of other animals. Humans who consume 'meat' are, perhaps unknowingly, colluding with the brutality of industrial farming. The other-than cash price such consumers pay is associated with morality. Vegetarianism, once connected with an ethical stance to the consumption of other animals, can now, through the drive of consumerism, be linked with individualistic lifestyles choices that emanate from concerns about personal health. Veganism has perhaps taken over the ethical role. But, whatever the reasons for vegetarianism, other animals would be eaten in greater numbers without vegetarians. Nevertheless, vegetarians and vegans are comparatively few in number; the consumption of other animals is an everyday occurrence for many and such consumption is central to leisure and culture, the subject of the next chapter.

7
Animals, Leisure and Culture

Introduction

On the BBC Radio 4 programme *A Point of View* (first broadcast on Friday 8 July 2011) Alain de Botton claimed that 'animals, as we know, don't loom very large in culture' (de Botton, 2011); he could not have been more wrong. If only, as a schoolchild, Alain had read Bryant's (1979) newly published paper about the zoological connection he would not have made such an error over 30 years later. Other animals are everywhere in culture; but like Mr de Botton we often fail to notice them. Perhaps the problem stems from our definitions of culture. As we use the word in everyday life, 'culture' has a number of meanings. We often experience 'different cultures' on a city break, we can be seen as having 'no culture' when we are uncouth, and we can be thought to be improving ourselves with 'a bit of culture' when we go to the theatre. For reasons discussed below, we associate 'culture' in all its forms with humans; however, other animals appear in all manifestations of 'culture'. For example, 'bullfighting' is associated with Spanish culture; accusing someone of talking 'bullshit' in a meeting would be considered to be coarse; and the tale in the opera Carmen takes place on the day of a bullfight. Although other animals are central to these notions of culture, they are often invisible within them. Moreover, other animals are conventionally not seen as having culture themselves. Culture is seen as exclusively human. Yet, culture is frequently associated with leisure and other animals are often central to our leisure activities. Whether we are eating other animal-based foods when dining in McDonald's, are riding a horse, are out hunting, are betting on a greyhound race, are being encouraged to purchase by cartoon animals in advertisements, are looking at other animals in zoos or are gazing at representations of other animals in paintings and films, we are engaging with other animals in some way. By drawing on insights associated with the sociology of culture and the sociology of leisure, in this chapter I explore the ways in which other animals are used by humans for the purposes of communication, relaxation, recreation and entertainment.

What is culture?

As we saw in Chapter 2, sociologists argue that nurture rather than nature is central to what makes us human. Culture is essential to nurture and, in consequence, the study of culture is central to sociology. Culture is used as a primary signifier of the difference between humans and other animals because, Mead (1934) argues, humans and humans alone have perception, imagination and language. Of course, more recent research has shown the error of such understandings. For example, Marc Bekoff (2002) uses the term 'minding animals' to refer not only to how we should care for other animals but also to how we should recognize that they have minds. Dogs who modify their behaviour in response to training (Sanders, 1999), cats who ritually greet humans (Alger and Alger, 1997), and birds and monkeys who use alarm calls to signal danger (Irvine, 2007) are some of the many examples of the minds of other animals (Chapter 2). Other animals are central to culture in all its forms and understandings, and there are a good number of understandings of what constitutes 'culture'.

Raymond Williams proposes that culture is one of the most complex words in English (1988, p. 87). A classic definition is put forward by Clifford Geertz who suggested, 'Believing, with Max Weber, that man is an animal suspended in webs of significance he himself has spun, I take culture to be those webs' (2003, p. 174). These webs span wide indeed. The term 'culture' is used in a range of discourses, both popular and academic; indeed, Peter Goodall comments that the word is so ubiquitous and is used in such diverse ways that 'I began to fear that the word "culture" might have entered a terminal phase of uselessness' (1995, p. xii). I do not have the space here to address the range of definitions or the range of problems; however, it is worth noting the important aspects as these will inform the discussion. Simon Stewart notes a tension between the 'restrictive usage' of the word and its 'general usage' (2010, p. 65), and we will see that other animals are important in both of these usages. Stewart points to Williams (1981), who contrasts the general usage of 'culture' as being the term as it is applied to a 'whole way of life' and the more restrictive application of the term that is associated with 'active cultivation of the mind' (2010, p. 65). Stewart proposes that 'Renaissance man' is the embodiment of the cultivated mind because Renaissance man is a polymath who is well-versed in, for example, literature, philosophy, science and music (2010, p. 65). Renaissance man is thus constructed as an earthly being who is as far from 'animal' as it is possible to be because, in his pursuit of a cultivated mind, he has left much of the natural behind. However, Stewart notes there is often a vacillation between this restrictive usage of the term and the general usage, which centres on a 'whole way of life' (2010, p. 65). He uses the example of French culture, which involves notions of 'a national cultural disposition incorporating ways of thinking, acting, working, eating and drinking' (Stewart, 2010, p. 65). Thus, we might identify eating snails as a part of French culture whereas roast 'beef' is seen as 'so English'. Importantly for the discussion here, whichever way we define culture, other animals are there in one way or another.

Sociologists emphasize that culture is not something that is simply 'out there'; we both act upon culture and are acted on by culture. Consequently, we are influenced by culture, but culture can be changed and new cultural forms and meanings can be generated. Queenie Dorothy Leavis (1978) draws distinctions between 'high' and 'low' culture, defining high culture as civilizing in its effects and low culture as having harmful effects on audiences. That which is defined as very low culture is severely admonished by society in general, not least because of the effects it is thought to have. For example, dog fighting was made illegal in the UK in part because it is thought to lead to further violence. In contrast, shooting grouse is viewed as a high-class 'sport' by some and is not considered to be the instigator of further violence in those who participate in it. However, activities can transgress the high/low culture boundary and sometimes this is prompted by those involved. For example, in a bid to preserve the once high-class 'sport' of hunting foxes with dogs, the poster campaign used by the UK pro-hunting group *The Countryside Alliance* tried to remove the 'toffs' association by presenting the activity as being for 'ordinary people' (Burridge, 2008, p. 31). On whichever side of the boundary a cultural activity is located, the idea of cultural effects is problematized by 'cultural studies', as this points to the interactive relationship between audiences and cultural forms. The field of cultural studies challenges the notion that people are merely passive 'dupes' who uncritically soak up the messages contained in cultural products, people rather are seen as audiences active in their consumption of culture, making decisions about what to watch, read or listen to, and about how to respond to the messages contained therein. For example, the Marxist approach taken by cultural studies in the UK (Alexander, 2003, p. 183) leads to an analysis of culture 'as a site of struggle' (Stewart, 2010, p. 65). In such studies the focus of analysis is removed from 'high culture' like paintings and literature to a focus on popular culture, which is seen as 'the lived culture of ordinary men and women' (Storey, 1993, p. 67). Rather than disparaging popular culture, the cultural studies approach celebrates it not least because it can be utilized by oppositional groups, marginalized groups and working-class groups (Alexander, 2003, p. 183). For example, The Smiths' song 'Meat is Murder' is devoted to oppositional messages about eating 'meat' and it is widely considered to be a 'vegan anthem' (Wallis, 2011, p. 8). So, audience reception of cultural products is central to the cultural studies approach. Hall's (1980) key work on encoding and decoding provides a valuable way of thinking about audience responses to cultural products. He distinguishes between the encoder (who is the creator of the product and its messages) and the decoder (the audience who views the product and defines the message). It is worth dwelling on an example. The person who creates a cultural product (let us say the film *Jaws*) encodes a meaning into that text (the message might be that humans must kill sharks who are out to kill them). Viewers must decode or read the messages in the text in order to make sense of it. Because members of audiences are active consumers, each member of the audience will not necessarily decode the message as intended by the creator and will not necessarily take away the same meaning as other members of the audience. Hall (1980) outlines four broad decoding positions that an audience member might take: the dominant-hegemonic position occurs when the viewer receives the intended message (thus humans must kill killer sharks); the oppositional position is taken when a viewer understands the intended message but chooses to take an alternative message (e.g. *Jaws* demonizes sharks who are only doing what comes naturally); the negotiated position sees a viewer mixing the first two approaches listed above (e.g. sharks are doing what they do naturally but humans should kill them if they come too close to human habitation); and the aberrant position finds the viewer interpreting the message in an unusual or peculiar way (e.g. the film is really about the cold war and the shark represents USA's fears about the Soviet threat). In consequence, we cannot merely assume that other members of the audience have received the messages that we think they have.

Thinking about other animals, Steve Baker agrees that audiences have different and often conflicting readings of cultural products, but the differences stretch wider than this as audiences have 'conflicting views over what constitutes "their" culture and their relation to or endorsement of that culture's dealings with animals. It is a matter, in other words, of the place and status of animals in relation to questions of identity' (1993, p. 25). This returns us to the more general use of the word 'culture', which refers to 'culture' as a 'whole way of life'; other animals are central to whole ways of life. Different cultures have varying cultural practices when it come to other animals, have different ideas about the general 'other animal' and have diverse views about specific other animals. Furthermore, within societies there are many cultural differences. Thus, in decoding cultural products an individual is 'not a free agent' who can change messages and read meanings at will (Baker, 1993, p. 26), rather the individual is somewhat constrained by culture as culture acts upon him or her. Because other animals are central to different cultures in similar and in different ways their positions and representations can be used to 'understand the meaning and basis of humanity' (Tester, 1991, p. 70). This brings me to the question 'why look at animals?'

Why look at other animals?

In his influential essay, John Berger (2009) begins to answer his question 'why look at animals?' with the observation that we have lost touch with other animals. He is interested in the interconnections between the living other animals from whom we are separated and the symbols of other animals with which we engage. He suggests that before the onset of corporate capitalism humans and other animals shared their spatial worlds both in an economic sense and in a productive sense (Berger, 2009, p. 12). However, Berger sees our present relations with other animals as being based in spatial separation. In this regard he is mainly centring on our relations with living animals (Chapter 5); humans continue to exist in spatial relation to, for example, dead animals through consumption (Chapter 6). Thus, as we saw in Chapter 5, in urban spaces human interactions with living other animals are expressed in the forms mainly associated with relationships with companions or 'pets', with viewing other animals as spectacles in zoos, or with trying to nurture wanted or exterminate unwanted 'wild' animals. Because we are so spatially separated from living other animals, Berger (2009) argues that we can no longer imagine them, and thus we are reduced to making representations of them in the forms of toys and pictures. Although Berger has been criticized for his romantic view of the past, the central aspect of his message is clear; human imaginings of other animals are central to cultures in which humans are separated from living other animals and in which other animals are invisible (Berger, 2009). In this vein, Baker invites the readers of his book *Picturing the Beast* to 'try to envisage the whole of the culture, the contemporary culture of which we are a part and then call to mind all the ways in which animals figure in its everyday operation' (1993, p. 3). He recognizes that this is not an easy task but, he suggests, if it could be done the list would have little to do with living other animals, rather it will be comprised of 'representations of one kind or another' (Baker, 1993, p. 3). This answers the question 'why look at animals?' Baker suggests that through such looking we are able to consider what the representations of other animals reveal about our attitudes to them and to humans, and we are able to examine the consequences of our attitudes (1993, p. 3).

When looking at other animals we need to understand the relationship between society and the representations at which we are looking. In order to think about this I draw on the work of Victoria D. Alexander (2003) who has explored the sociology of the arts, a study that is very useful to us here. Although Alexander advocates a more complex approach to understanding the relationship between the arts and society, I intend to focus on two clear approaches that she discusses, shaping and reflection approaches. Reflection approaches are multifarious; however, broadly such approaches maintain that art contains information about society (Alexander, 2003, p. 21). So, if we want to learn about the position of other animals in a given society, and the relations between humans and other animals in that society, we might watch popular cultural forms (likes films or television programmes) to see how other animals are portrayed and how humans are portrayed in relation to other animals. The obverse of this approach is the shaping approach, which focuses attention on how art-forms influence our behaviour and the ways in which we see phenomena (Alexander, 2003, p. 41). Again this approach encompasses a wide group of theories, but all share the core belief that art has an effect on society. For the most part shaping approaches have centred on the negative effects of cultural forms, such as the view that violent television drama can induce violence in children who view it (Alexander, 2003, p. 53). So, films that contain violence towards other animals might have the effects of legitimizing animal cruelty and thus lead to acts of cruelty to other animals. However, studies also recognize that art can shape in a positive way. For example 'green art' and documentaries about other animals might encourage an understanding of other animals and a concern about conserving their habitats. It seems quite obvious that each of these effects cannot be studied in isolation – the arts both reflect and shape societal beliefs about humans and other animals. As Baker argues, 'Culture shapes our reading of animals just as much as animals shape our reading of culture' (1993, p. 4) and, I think, our reading of culture shapes how we see other animals.

In order to 'look' at other animals we must look at our own culture, but, of course, this is not as easy as it sounds as we are so embedded in it. Baker proposes that a study of our own culture should draw on 'mentalités', which Derrida (1982, p. 135) defines as 'the cultural analysis of popular behaviour and attitudes' (1993, p. 6). Baker draws on Robert Darnton's (1985) explanation that makes clear that drawing on mentalités means 'treating our own civilization in the same way that anthropologists study alien cultures' (Baker, 1993, p. 6). So we might ask, following Baker, 'What place does the animal hold in our imagination, and how are we to understand the uses to which our imaginative conception of the animal is put?' (1993, p. 6). Because Baker is interested in the preconceptions and biases that are alive in us now, he proposes that the mentalité is the present (1993, p. 7), which makes it eminently useful for a sociological account. But this application to the present leads to questions that, Baker notes, centre around the ways in which the present mentalité might be inaccessible to us, and how we might deal with any inaccessibility (1993, pp. 7–8). In this regard, we need to pay particular attention to the reasons why other animals have become invisible. Referring to the notion of 'naturalisation' as it is used by Roland Barthes (1993), Baker notes that phenomena like other animals are not really hidden, rather 'culture typically deflects our attention from these things, and makes them seem unworthy of analysis' (1993, p. 8). We will come to this later. Firstly, I want to explore one of the ways in which sociology has traditionally looked at other animals, which centres on the use of representations of other animals as ritualistic symbols.

Sociology and other animals as ritualistic symbols

Other animals have made a major appearance in sociological analysis in their representations as ritualistic symbols. Durkheim, for one, was very interested in the ways in which we arrange all 'real and ideal' things as sacred or profane (1965 (orig 1915), p. 37). He argued that we organize our surroundings by defining most phenomena as profane, that is as ordinary phenomena of everyday life. However, we set some things apart, designating them as sacred and this, he argues, is characteristic of religious thought (Durkheim, 1965 (orig 1915), p. 37). Designation as sacred is not confined to 'personal beings that are called gods and spirits; a rock, a tree, a spring, a pebble, a piece of wood, a house, in a word anything can be sacred' (Durkheim, 1965 (orig 1915), p. 37), and the origins of such designations, and of our own religious beliefs, are found in totemism. The totem is an expression of something that is sacred and it often takes the form of a representation of an other animal. However, the focus is not on the other animal as such, rather '...the totem is before all a symbol, a material expression of something else' (Durkheim, 1965 (orig 1915), p. 206). In this way, the totem is a 'visible or outward form' of the god and of the society in which it is sacred (Durkheim, 1965 (orig 1915), p. 206), and totems, like all religious forms, enable people to express their social unity. Totems garner huge respect. Unlike the living other animals whose image they take, the totem can only be touched by certain members of society (Durkheim, 1965 (orig 1915), p. 133). Most importantly, the totem represents the 'the moral life' of the society (Durkheim, 1965 (orig 1915), p. 190) and thus the totemic animal provides the moral authority to a society without the need for coercion (Durkheim, 1965 (orig 1915), p. xvi). In this regard 'the totem is their rallying sign' and members of a society 'become conscious of the kinship uniting them' (Durkheim, 1965 (orig 1915), p. 358). But why are so many totems comprised of representations of other animals? In attempting to answer this question Levi-Strauss rejects Durkheim's view of totemism, as he feels that Durkheim has reduced the totem to something sentimental (1962, p. 102). Levi-Strauss is interested in the capacity of the totem to express structures of difference. The totem is a symbol that is used by one group to classify itself as different from another and 'the animals in totemism cease to be solely or principally creatures which are feared, admired or envied; their perceptible reality permits the embodiment of ideas and relations' (Levi-Strauss, 1962, p. 162). So, Levi-Strauss concludes, totemic symbols serve to distance humans from other animals. Still, other animals are chosen as totems because 'they are good to think' (1962, p. 162), as totemic representations of other animals are used as expressions of features of human life deemed important by the societies that use them. The process of differentiation from other animals is not limited to the use of totemic symbols, we can find such processes in 'fine' and 'popular' cultural forms such as paintings and sculpture, and television and film, and also in the ways in which other animals are used for spectacle and sport.

Representing humans and other animals

Cultural forms often embody ideas about human distinction from other animals. Human cultures are more often than not grounded in the notion that humans are very different from other animals; it is usually only 'faulty' humans or humans who have transgressed boundaries in some way or other who are represented with 'animal-like' characteristics. Thus, cultural representations are extremely important in reflecting back to us our own notions of our human identity.

Reflecting the human: Notions of human–animal difference

Humans have spent a good deal of time distinguishing themselves from other animals. As Levi-Strauss observes, 'It is because man originally felt himself identical to all those like him (among which, as Rousseau explicitly says, we must include animals) that he came to acquire the capacity to distinguish *himself* as he distinguishes *them* – i.e. to use diversity of species for conceptual support for social differentiation' (in Berger, 2009, p. 17). Representations of other animals reflect and shape our notions of difference between us and other animals and, in doing so, reflect and shape how we see ourselves as human. Thinking about image-based representations alone, Baker (1993) notes the astonishing complexity and diversity of images of other animals. Instead of taking the images for granted, in problematizing the images Baker sees that there is much more going on than is immediately apparent and this depth of looking indicates societal notions about what it is that makes us human. In order to explore such notions we must study a mentalité (1993, p. 6). Accordingly, asks Baker, 'why is it that our ideas of the animal – perhaps more than any other set of ideas – are the ones which enable us to frame and express ideas about *human* identity' (1993, p. 6). The ideas are not only manifest in images; metaphors using other animals are ubiquitous in various cultural products. However, in whichever form such ideas come, they are often designed to distance the human from the 'animal'.

Thinking back to the 'in' identity of 'human' and the 'otherness' of all other animals (Chapter 3), here I aim to explore how cultural artefacts such as language, paintings and literature serve to express our ideas about human difference from other animals. In order to demonstrate this difference, cultural artefacts and products often have to draw on other animal imagery and metaphors that invoke other animals and this is most obvious in the ways in which the supposed characteristics of other animals are summoned to reflect human faults (Kemp, 2007, p. 34). This is most clear in language that denigrates a 'human' by calling him or her an 'animal'. We see this in the news media almost daily. Perpetrators of the most heinous crimes are often referred to as animals (Chapter 4) and the most serious charge that can be laid against humans is that they have behaved like 'animals' (Midgley, 2004). Consequently, to call a human an 'animal' is to reduce their status (Kemp, 2007, pp. 34–5). However, Martin Kemp notes, 'When it comes to the generic use of the term "animal" in a perjorative sense, we are invariably unfair to the animal kingdom' (2007, p. 35). There are no serial killers, rapists or suicide bombers among other animals. Interestingly, Kemp observes that the 'science of physiognomy' is to blame for this denigration of other animals. This returns us to the ideas of Lombroso, who we thought we had left behind in Chapter 4 and in the nineteenth century. Lombroso argued that criminals have distinctive characteristics such as excessive hairiness, prominent jaws and relatively long arms (Gould, 1980). Kemp argues that the notion that 'we can identify the appearance and behaviour of particular individuals or human types with animals has manifested itself across remarkably wide fields of cultural expression over the centuries' (2007, p. 35). He cites the example of the fifteenth-century Dutch painter Hieronymus Bosch, who felt that the characteristics of different human personalities were inscribed on the features of their faces. He put these characteristics in his pictures as they are 'prescriptions that the artist could rely upon the spectators to recognize' (Kemp, 2007, p. 35). Thus, in his picture 'Christ Taken Captive' the soldiers are given bestial faces which contrast starkly with the gentle face of Jesus (Kemp, 2007, p. 34). This is not something that only happened in the distant past when Bosch was making his pictures; such imagery is still used when we want to denigrate humans (Kemp, 2007, p. 41). The characteristics we use are often drawn from the features of other animals and the artworks that use them frequently have a humorous dimension (Kemp, 2007, p. 36). For example, Kemp notes that we cannot distinguish the human from the pig by the time we get to the end of George Orwell's 1945 book *Animal Farm* (2007, p. 36). This merging is seen in the 2007 *Keep Britain Tidy* 'Dirty Pig' campaign, in which humans who drop litter take on pig characteristics (the man a pig's nose and the woman a pig's tail). Such representations serve to distance these 'bad' humans, who take on the characteristics of other animals, from 'good' humans, who do not.

Scientific theories about the distinctiveness of humans in relation to other animals have had an important effect on cultural representations of other animals. As we saw in Chapter 2, Descartes saw a natural distinction between humans (as capable of thought) and other animals (as machines), and this sharpened the differentiation between humans and other animals. Although the grounds for Descartes' unsophisticated mechanistic belief that animals are actually unconscious has been undermined (Midgley, 2002, p. 138), still our current thinking about the fundamental nature of the differences between humans and 'animals' stem from Descartes' ideas (Dupre, 2002) (Chapter 2). These differences continue to be played out in cultural representations of humans and other animals, and it is the credence given to the assumptions of difference that puts the comic into images of humans who are given the characteristics of other animals. The image of Charles Darwin as a monkey is a case in point; and we need to apply the same idea of absolute difference in order to get the joke when the then USA President George W. Bush was depicted as a monkey and when the then UK Prime Minister Tony Blair, once depicted as Bambi, ends up being depicted as a poodle. However, Darwin's ideas about evolution resulted in a redefinition of the hard distinction between humans and other animals (Kemp, 2007, p. 39) and this is most obvious in cultural representations of human evolution that begin with a crouching nonhuman primate and end with a human standing upright (or more recently, with a human sitting at a computer). But even before Darwin's theory of evolution became public, notions of gradations of difference in intelligence between 'human' and 'animal' had gained ground (Kemp, 2007, p. 39). In consequence, racist notions of differences among humans were reflected in cultural representations. As an example Kemp points to Charles White's account of 'the regular gradation in Man, and in Different Animals', which provided an elaborate visual schema of abhorrent and preposterous notions about the inherited superiority of white humans compared with others (Kemp, 2007, p. 39). This 'gradation' blurs the hard break between 'human' and 'animal'; the break depicted as a progressive development that leads to the white human. Darwin's theory of evolution, which transformed the idea of the unchanging distinctiveness of humans from all other creatures, served to cement 'gradation' ideas still further because 'Once one species could potentially flow into another over eons of deep historical time, the relationship between humans and animals needed to be redefined' (Kemp, 2007, p. 39). This redefinition has served odious and ridiculous racist depictions of black men as more ape-like than white men (Kemp, 2007, p. 40). These 'simianized images' (Baker, 1993, p. 115) are used to depict hateful messages about specified groups. Although such imagery has become less acceptable in a range of media, such preposterous and abhorrent images might still appear in racist propaganda (Baker, 1993, p. 115).

Representing women and animals

In the famous anti-fur poster that shows a high-heeled woman dragging a fur coat that leaves a bloody trail, the caption reads 'It takes up to 40 dumb animals to make a fur coat. But only one to wear it'. The image is shocking; it is an important reminder of the horrors of the fur industry. Wearing the fur of particular other animals is associated with luxury in a number of countries and the industry involves both trapping and farming of other animals for their fur. In the fur farming industry alone, 50 million minks and 7 million foxes are bred for their fur each year (Linzey, 2009, p. 97). Because the industry demands that the skins are undamaged the ways in which these other animals are slaughtered results in terrible suffering (Linzey, 2009, p. 98). Even though fashion houses such as 'Gaultier' and 'Armani' have attempted to revive fur as a desirable material for clothing, wearing fur remains abject in many cultures. Campaigns against the wearing of fur have been very successful; however, referring to the campaign poster mentioned above, Adams argues that the imagery used by campaigns has at times 'reinforced bias that belittles women' (1995, p. 134). Although wearing fur has become repulsive in many countries, still the notion that women wear fur coats 'is embedded in malestream culture's image of femininity' (Adams, 1995, p. 134). This malestream point of view, which gives no concern for gender inequalities, is propped up by the 'arrogant eye' (Adams, 1995, p. 134). The concept of the 'gaze', as is it is used in sociology, is useful to us here as it enables a consideration of how viewers think about that which is presented and the effects on the subject (which becomes an 'object') of the gaze. The concept is grounded in Foucault's discussion about the 'medical gaze' (1976, p. 89), which centres on the way in which the patient is dehumanized through the separation of their body (the object of the gaze) from their identity. The concept of 'the gaze' as it is used in analyses of culture encourages consideration of how an audience views that which is presented to them. In this way 'the gaze' as a concept does not have to be confined to the systematic and scientific gaze of a medical professional in an institution, it can be applied to us as we look upon the world (Urry, 1990, p. 1). Gazes are often constructed through difference, such as by society and by a social group (Urry, 1990, p. 1). The concept is employed in feminist theory (e.g. Adams, 1995) where it can be used to explore how men look at women, how women look at themselves and other women, and the effects of this looking. This is important as the gaze can reflect the structures of power or the nature of a relationship between the subjects, because although it might appear that it is only about 'looking' the gaze actually denotes a power relationship 'in which the gazer is superior to the object of the gaze' (Schroeder, 1998, p. 208). Accordingly the 'male gaze' is where the male imposes an unwanted gaze upon the female; this imposition underlines the power of the male gaze. In such looking the subject (in this case the woman) is objectified and this objectification means that the woman becomes a representation. Adams explains that a main way of 'making a subject into an object' is by making depictions or representations of them (1995, p. 41). In many cultures women and other animals are looked at as objects and, Adams suggests, because women and men 'assimilate patriarchal culture; the human male gaze is exhibited by both when looking at other animals' (1995, p. 41). Adams uses the concept of the 'arrogant eye' to signify the process of the objectification of beings. This 'arrogant eye' is evident in the coupling of femininity with the wearing of a fur coat (Adams, 1995, p. 134). Advertisements for fur coats depict the women to be gazed upon, and so do images in anti-fur campaigns (Adams, 1995, p. 134). Because femininity is shaped around fur, Adams grants that the majority of fur coats are worn by women but, she argues, the image-based element of the campaign is dressed in high heels, has taken off the fur coat and is dragging it along the ground in a sexually provocative pose. Of course, this representation could not be of a man because, in our culture, it is women who are culturally aligned with fur and it is usually women who are portrayed as sexual provocateurs. In addition to that which is contained in the image, Adams is critical of the caption. Identifying women as 'dumb animals' is evocative of 'the Western political tradition, that has viewed women precisely as dumb animals' and for this reason alone it is clear that the headless person could not possibly be a man (1995, p. 135). Thus, Adams concludes, this anti-fur campaign is colluding with an objectification process as the poster reflects a patriarchal culture that objectifies women and other animals (1995, p. 135). This objectification of women and other animals is apparent in other cultural forms, for instance in sport.

To emphasize his observation that sport is central to contemporary culture, Barry Smart cites Phil Knight's (the founder of Nike) remark that sport seems set to 'define the culture of the world' (from Katz, 1994 in Smart, 2005, p. 3). Smart argues that 'sport is recognised to be one of the key cultural institutions involved in the constitution of national identity' (2005, p. 3). Foxhunting in Britain; grouse shooting in Scotland; hunting, fishing and shooting in the USA; bullfighting in Spain; and horse-racing in Ireland demonstrate how central sports that involve other animals are to national identities. Moreover, these 'sports' involve cruelty to, and sometimes the killing of, other animals, and this seems to contradict legislation in some countries that is increasingly designed to protect other animals against cruelty and brutality (Franklin, 1999, p. 105). Often such brutality is associated with gender. For example, although there have been increasing numbers of women in Spain involved in bullfighting (Pink, 1997), still killing the bull is associated with masculinity (Douglass, 1997). The bull is viewed as the 'quintessential male animal, its aggressive male qualities...being associated with great sexual potency' (Marvin, 1988, p. 40). However, the challenge between the human male and the bull is only made possible because the 'bull is forced into being in the position of being "out of place", and the death symbolizes the "correct" relationship between man and other animal and between the "civilized" and "nature" ' (Marvin, 1988, p. 40). Nevertheless, bullfighting has had to adapt to the changing social context which, though contested, has led to some success for women bullfighters (Pink, 1997).

The gendered character of bullfighting is apparent in other such sports. In an analysis of a random sample of 15 issues of the sport magazine *Traditional Bowhunter* (published between 1992 and 2003), Linda Kalof, Amy Fitzgerald and Lori Baralt (2004) sought to investigate the juxtaposition of hunting, sex, women and other animals by decoding the photographs, narratives and advertisements used in the magazines. The magazine currently parades the mantra 'Progress does not necessarily mean letting go of tradition' (Traditional Bowhunter, 2011), and this tradition is hunting other animals with bows and arrows. The first edition of the magazine appeared in 1989, and when Kalof et al. undertook their research, the magazine had 60,000 readers in seven countries (Kalof, Fitzgerald and Baralt, 2004, p. 237). Kalof et al.'s analysis revealed that the hunting discourse central to the magazine is grounded in 'the sexualization of animals, women, and weapons, as if the three are interchangeable sexual bodies in narratives of traditional masculinity' (2004, p. 237). Sexualized representations of women and other animals in the magazines often drew on stereotypical feminine characteristics associated with 'patriarchal versions of romance'. For example, they found turkeys referred to as 'redheads'; a 'decoy' referred to as 'Barbie Hen' and deer antlers referred to as 'big'uns' (Kalof, Fitzgerald and Baralt 2004, p. 242). Such sexualized symbolism of women and other animals is summed up in the following extract from the magazine (reproduced in Kalof, Fitzgerald and Baralt):

> Developing my sense of smell had one unexpected consequence, particularly while I was in college. I would return to campus after several days of camping and hunting to find that the scent of those college girls was, to say the least, an added distraction from my studies. The experience gave me insight into the state of mind of a buck deer during the rutting season. Just the sight and sound of coy young does everywhere is enough to cause madness but add their scent and it might be enough to cause a bull to run to the nearest hunter and say, 'Just shoot me'!
>
> (2004, pp. 241–2)

Of course, such discourses are not confined to the *Traditional Bowhunter*, if only they were. Kalof et al. observe that popular cultural forms such as 'pornography, music videos, prime-time television, feature-length films, magazine advertisements, and narratives of black slavery' associate hunting other animals with sex and link women with other animals (2004, p. 238). The narratives in such cultural products construct and reproduce relations of domination, such as patriarchal relations, in society (Kalof, Fitzgerald and Baralt, 2004, p. 239). A further aspect of this is that 'the cultural construction of hunting is rooted in a symbolic system that values predation and dominance and conjoins hunting and sex with women and animals' (Kalof, Fitzgerald and Baralt, 2004, p. 239). Citing Collard, the writers take the point that our cultural mentality is associated with predation; for example, we hunt for jobs, chase the stock market and hunt out consumer products (Kalof, Fitzgerald and Baralt, 2004, p. 239). Thus, recalling Adams, they note a 'logic of domination' that underscores discourses associated with hunting (Kalof, Fitzgerald and Baralt, 2004, p. 239). This logic of domination is found in the anthropomorphization of living other animals, which is also gendered. For example,

> When alive and being chased in the sport of hunting, animals are given human characteristics (primarily feminine), but when dead and displayed as a trophy, anthropomorphism is no longer necessary, humans are distanced from the animal, and the animal is simply dead.
>
> (Kalof, Fitzgerald and Baralt, 2004, p. 246)

Kalof et al. conclude that popular cultural images like the ones in *Traditional Bowhunter* 'celebrate and glorify weapons, killing, and violence, laying the groundwork for the perpetuation of attitudes of domination, power, and control over others' (2004, p. 247).

The above discussion points to the ways in which, in culture, other animals reflect, not least, human identity, patriarchy, masculine identity, constructions of femininity and the construction of human sexuality. Although the other animal is present, the focus is on the human. However, if we try to decentre the human subject we will see the other animals more readily, because 'the decentring of the human subject opens up a valuable conceptual space for shifting the animals out from the cultural margins' (Baker, 1993, p. 26). In order to do this I want to move away from what human representations of other animals say about us and move towards our imaginings of other animals.

Representing other animals

Baker presents a number of clues as to why representations of other animals matter. One of his areas of interest is the representation of other animals in images associated with conservation. He refers to Tim Ingold's (1994) deliberations on images of other animals in totemism and in conservation in which Ingold ponders over the relationship between imagined other animals and their real counterparts. Ingold asks 'Do animals exist for us as meaningful entities only insofar as each may be thought to manifest or exemplify an ideal type constituted within the set of symbolic values making up the "folk taxonomy" specific to our culture?' (1994, p. 12). Baker responds that we can only ever know other animals in their mediated forms; we cannot know them directly. In his words, 'To see animals at all is to see them as something – as something we have made meaningful' (Baker, 1993, p. 180). So the 'real' is inseparable from the representation and most importantly for Baker, in the field of conservation, we may have a greater investment in the representation than we care to admit (1993, p. 178). Baker cites as an example the World Wildlife Fund for Nature (WWF) logo of the panda. He recalls that it has been 'unkindly suggested' that the WWF has donated a great deal of money to preserve the panda because it has 'staked its own visual identity by choosing it to adorn the WWF letterhead' (1993, p. 179). In this way the panda is the WWF's 'totemic animal' via which we try to make sense of the other animals we want to conserve (Baker, 1993, p. 178). Humans are in a paradoxical relationship with living pandas and the panda logo is emblematic of this inconsistency; humans are a threat to the species and yet they suggest that they hold the key to the safety of the species (Baker, 1993). Accordingly, the way in which we represent and see the panda is mediated in that it is a reflection of our own concerns.

Disnification

Baker has spent much time looking at and problematizing images of other animals. His research reveals that representations of other animals are almost always funny, or are at least likely to be construed as humorous (1993, p. 23). Such representations make a 'nonsense of the animal' and, as a result, 'the animal is the sign of all that is taken not-very-seriously in contemporary culture: the sign of that which doesn't really matter' (Baker, 1993, p. 174). Baker's term for this frequent trivialization and deprecation of other animals is 'the disnification of the animal' (1993, p. 174). Disnification is explicitly image based and it serves to 'render [the other animal] stupid by rendering it visual' (Baker, 1993, p. 174). However, Baker does not centre his analysis on or reserve this term for the images of other animals that are used in Disney productions, rather he explores the everyday ways in which terms like 'Mickey Mouse', which evokes the cartoon image, are used descriptively to trivialize. In this way, instead of making sense of an other animal a disnified image is making nonsense of another animal. Consequently a disnified image contains an inherent problem because the image tends towards a stereotypical representation rather than the 'rigour that might apply in other nonvisual texts' (Baker, 1993, p. 175). Such images are used within popular and 'fine' cultural forms, including still photography, film, news media, 'animal advocacy' publications and adverts for 'animal-free' consumer products.

To provide an example of the disnification process Baker draws our attention to the pre-Disney forms of traditional fairy tales. Traditional fairy tales were originally intended to be part of a child's maturation process as they encouraged consideration of conflict. Other animals who appeared in these tales corresponded with this original conception. Thus, the wolf who appears in Little Red Riding Hood would be part of the wilderness, which was treated as a frightening place in conflict with the human. Of course, such imagery serves to distinguish humans from the 'natural world'; however the Disney version trades the dangerous wolf for a ridiculous wolf. In the original Disney version the wolf is a cartoon character who wears a top hat, white gloves and patched red trousers held up by green braces. In such representations Baker notes that the Disney Corporation seems to have failed to appreciate the true meaning of the fairy tales and 'has settled instead for a trivializing and sanitized cuteness which misses out on (or sometimes even contradicts) their mythological richness and their psychological depth' (1993, p. 177). Thus, Baker contends, disnification is not concerned with the original meaning of these tales and thus the representations of the other animals who appear in the disnified versions should be 'approached and understood on this basis' (Baker, 1993, pp. 177–8). The disnified reduction of other animals to silly, trivial and stupid serves to accentuate human superiority. This is especially so in films in which other animals talk. For example, in the film *Ratatouille* the much vilified rat becomes the cuisine loving Remy, a rat who wants to be a French chef. In the 2008 DreamWorks film *Kung Fu Panda,* a panda who is a novice at martial arts becomes chosen to be the Dragon Warrior. This recalls the WWF logo of the panda. Baker observes that the logo is 'pleasure-inducing' and cuddly like the toy teddy bear (1993, p. 182). He proposes that this neotenized form (i.e. a form that is given immature or infantile features when an adult) in no way resembles the real panda. Drawing on work of Bob Mullan and Garry Marvin (1987) on the function of the panda logo, Baker remarks that unintentionally the logo reveals the essentially human-centred project of conservation, a project that is really about conserving chosen other animals for human pleasure (1993, p. 182).

Looking at other animals

What do we think we know of other animals when looking at living animals? We might look upon them on the race track, in the bull ring, in the circus, in the zoo or in a television wildlife documentary, and when we look upon them in these places we think we are seeing the 'real thing'; we think we can distinguish the other animals we look upon from the depictions talked about by Baker. We have already taken some note of what our spectatorship of living other animals says about us but here I want to centre on the other animals. A most obvious place in which we look upon living other animals is in zoos. For Malamud, 'zoos are one response to people's need and desire to know other animals' (1998, p. 257); but does this enable the spectator to 'know' the other animals? Let us return to Berger. Berger views the cages in zoos as being like picture frames that encase other animals inside them (2009, p. 33). Visitors to zoos move from one cage to another, rather like they move from one picture to another in a gallery, to look at the other animals in the frames. But, Berger maintains, 'the view in the zoo is always wrong' because 'you are looking at something that has been rendered absolutely marginal' (2009, pp. 33–4). But, we might protest: many zoos no longer keep other animals captive in such cages; the anthropocentric representations of caged other animals in zoos, which were so notorious in the 1970s, have given way to innovative places in which other animals live in simulations of the complete ecosystems from which they came (Franklin, 1999). Nevertheless these other animals, like their caged forbears, are only free within limits (Berger, 2009, p. 34). The backdrops and places are merely 'tokens' that serve to encourage the viewer to think they are seeing the 'real' thing. In such enclosures other animals are often isolated from each other, they rely on humans to feed them and thus their responses to other creatures, and to their environment, is necessarily changed (Berger, 2009, p. 35). Paradoxically, although captive, Berger agrees that the keeping of 'specimens' of other animals in zoos might guarantee a prolonged existence to the species that would otherwise be threatened if they were free. As we have seen (Chapter 5), zoos often describe their purpose as the protection of other animals who are members of endangered or rare species and/or the promotion of the repopulation of wild habitats (Franklin, 1999). Indeed, this human response is understandable, given that 'the number of species endangered because of man takes one's breath away' (Derrida, 2004, p. 120). However, keeping other animals in captivity in zoos merely reflects our own concerns about which other animals we want to conserve and the view of nature that we want to maintain (Mullan and Marvin, 1987). But unhappily, Malamud contends, 'zoos have historically demonstrated their unconcern, or incompetence, with regard to achieving any beneficial impact upon the natural world outside their gates and cages' (1998, p. 46). As we saw in Chapter 5, Malamud asks 'Does an ocelot really enjoy the passing crowds looking at it all day?' (1998, p. 47).

What does the ocelot think of our gaze? Malamud's question focuses our mind on the other animal as a being who is looking at us. Because we constantly do the looking, and construct ourselves as the see-ers (the arrogant eyes) we forget that the other animal is looking back at us. This returns us to Derrida. Upon coming out of the shower in his apartment, Derrida finds that he is embarrassed by his nudity because of the gaze of the cat who lives with him (2004, p. 120). Derrida realizes that 'his cat' is not just an 'an exemplar of the species' but is a specific 'unreplaceable living being' (2004, p. 120). When we are looking at other animals they are looking back at us. In looking at us looking at them, Berger argues, 'The animal scrutinizes [humans] across a narrow abyss of non-comprehension... The man too is looking across a similar, but not identical abyss of non-comprehension. And this is so wherever he looks. He is always looking across ignorance and fear' (2009, pp. 13–14). Berger concludes that the other animal is isolated and 'as for the crowds, they belong to a species which has at last been isolated' (2009, p. 37); thus we do not see the other animals at all.

Concluding remarks

Alain de Botton was clearly wrong in his suggestion that other animals are absent from culture; other animals are everywhere. They appear in 'culture' as it is applied to a 'whole way of life' (e.g. in ideas that associate the British with 'beef') and in the more restrictive application of the term that is associated with the 'active cultivation of the mind' (e.g. in the paintings by Hieronymus Bosch). In both cultural applications, other animals appear as representations of what we feel we are not and thus are important in our notions of our human identity. Humans who transgress are often depicted as more animal-like. Such cultural representations have been used in hateful ways to depict odious ideas about the 'otherness' of groups of humans. In addition, they have been used to denigrate women and other animals by making links between sex, women and other animals. All such representations reveal that human identity is based on notions of the 'natural superiority' of at least some (mainly white male) humans. Because our interactions with living other animals are increasingly diminished and lacking, we rely ever more on representations of other animals. However, these representations are not the 'real things'; they are merely imaginings that often serve to denigrate or disnify other animals. Even though we might become aware of this, we can only see 'real' other animals through these representations. This creates problems for human advocates of other animals because, Baker (1993) argues, until other animals are disentangled from the baggage of these images and representations of them as imposed by humans 'neither they nor their predicaments can be clearly seen'. 'Animal advocacy' is the subject of the next chapter.

8
Animal Experiments and Animal Rights

Introduction

This chapter explores grass-roots activism as it is associated with other animals in society. Grass-roots activism works at the level of subpolitics (Beck, 1992) rather than at the level of political parties. Grass-roots groups, such as Greenpeace, the National Anti-Vivisection Society (NAVS) and People for the Ethical Treatment of Animals (PETA), endeavour to generate public support for campaigns against the ill-treatment of animals. Other grass-roots groups, such as Pro-Test, try to generate public support for the use of animals when it is beneficial to humans (in this case for the use of other animals in experiments). Sociology has a good deal to say about such subpolitical expression. Using sociological perspectives on social movements that organize around specific goals, this chapter explores the ways in which humans have mobilized around issues associated with the human (ab)use of other animals. Although I draw on a number of issues, the main focus of the chapter is on the contentious issue of experiments on other animals. Experiments on other animals are a major focus of political engagement, with campaigning groups advocating for and against such actions. The chapter provides a more in-depth discussion of animal advocacy than that provided in Chapter 4 and draws out some of the policy issues associated with such experiments. Furthermore, the issue of experiments on other animals draws us back to the notion of the objectivity of sociology, which will be the main focus of the final chapter of this book. Before I turn to philosophical positions on the advocacy of other animals (which is a major focus of grass-roots activism) I want to start with a brief discussion about experiments on other animals. In order to do this I utilize the sociology of risk, as ideas about risks to human health have been most often used to justify experiments on other animals.

Experiments on other animals

Annually an average of 115 million living vertebrate other animals are used in experiments worldwide (Taylor, Gordon, Langley and Higgins, 2008, p. 327). This staggering figure can only be estimated as the majority of countries (79%) do not publish details of the number of other animals used (Taylor, Gordon, Langley and Higgins, 2008, p. 327). However, using statistical calculations Katy Taylor and her colleagues estimate that the USA, Japan, China, Australia, France, Canada, the UK, Germany, Taiwan and Brazil use the highest number of other animals in experiments (2008, p. 327). In 2005 in the European Union (EU) alone, 12.2 million vertebrate living other animals were used in experiments (Commission of the European Communities, 2007), and in 2010 3.6 million other animals were used in experiments in the UK, representing a 3 per cent increase on the figure for the previous year (Home Office, 2011, p. 8). These figures only cover mainly vertebrate other animals; there are many additional invertebrate other animals who are not included in the statistics as they are not considered to be 'animals' for the purposes of 'protection' (Peggs, 2010) (see below).

The use of other animals in experiments is one element of the overall project of human progress. For hundreds of years other animals have been treated as resources for human progress (Franklin, 1999), for example in the reproduction and manipulation of other animals for human food (Chapter 6), and none more so than in the advancements that are associated with biomedical research designed for human health benefits. To be sure, experiments on other animals do not solely focus on biomedical research as globally other animals are used to determine the effects and safety of a range of products, processes and manipulations associated with warfare, accidental damage, food processing, cosmetic enhancements and genetics (see Grant, 2006). To use the EU as an example, permitted purposes for such experiments are the development, manufacture and testing of drugs, foodstuffs and other products; disease prevention, diagnosis or treatment; assessment, discovery, regulation or modification of physiological conditions in humans, other animals and plants; and the protection of the natural environment (Council of the European Communities, 1986, Articles 3a and 3b) (for further discussion, see Peggs, 2010). Using other animals in any form of experiment is usually controversial. Here I concentrate on biomedical experiments on other animals as, given the focus on human health, this usually receives more public support than other forms. For example, a 2006 UK YouGov Poll found that 70 per cent of respondents felt it was acceptable to test new medical treatments on other animals compared with 18 per cent of respondents who thought it was unacceptable under any circumstances (YouGov, 2006). In order to gain some insights into the possible reasons for such support I want to return to risk and society as it applies to the issue of human health.

Risk and human health

Because our everyday lives are so infused with risk (Tulloch and Lupton, 2003), Piet Strydom marks out risk as the 'signature of contemporary society' (2002, p. 4). Beck argues that risk has such a high profile in our lives that it, rather than wealth, is the main principle around which inequalities are organized (Beck, 1992, p. 19). We are so concerned about risk that we no longer centre on trying to acquire good things, instead we focus much of our attention on trying to prevent the worst (Beck, 1992, p. 49). Thus, Bauman observes, we live our lives in perpetual anxiety (2006, p. 11). Is it any wonder we get ill? In order to make our anxious lives more bearable we try to predict and control an assortment of hazards (Bauman, 2006, p. 6), especially those associated with human health.

Every day the news media inform us about a range of human health risks, many of which are associated with other animals. Some media reports centre on risks to our health from living other animals, like those from 'dangerous dogs' (Chapter 4); others focus on the risks associated with consuming dead other animals, like CJD (Chapter 6); and still others simply invoke other animals, such as media reports about the risk of an epidemic of 'avian flu' or 'swine flu'. Concerns about diseases that we might 'catch' from other animals have been a major concern for many years, but this concern is only one part of a whole gamut of threats associated with human health. For example, media reports centre on the risks of heart disease from eating too much fat; food poisoning from eating certain salad products; skin cancer from getting too much sun; vitamin D deficiency from getting too little sun; the side effects associated with a specific pharmaceutical; the side effects of not taking the same pharmaceutical. Thus, we feel increasingly menaced by a range of health hazards, even though our overall health has generally improved (Busfield, 2006, p. 299) and we are, by and large, more free from serious diseases than we were in the past (Giddens, 1991, p. 115). Where are we to turn? Experts are central to our attempts to manage risks

(Giddens, 1991). Technological and scientific research points the way to a possible utopian future based in the human ability to prevail over our bodily limitations (Turner, 2007); however, this utopia is founded in a past and present in which millions of other animals have been commodified (Chapter 6), and this looks set to continue into the foreseeable future (Peggs, 2011a, p. 51). Other animals used in experiments are treated as products just like those who are used for human food and other consumables (Chapter 6), and in being so treated they are not only seen as the harbingers of human health risks, they are also seen as the commodities who can help us fight against these risks. Regarding past and present commodifications of other animals for human health gains, the pressure group 'Understanding Animal Research' (2011) proclaims a range of medical advances, including the development of insulin, penicillin and vaccines, that have relied on experiments using other animals and a range of scientists claim further possibilities associated with, for example, genetics (see Peggs, 2011a, p. 51). Other animals who are used in biomedical experiments are usually 'standardized animals' who have been bred for specific biomedical purposes, and the commodification is so advanced that humans actually patent other animals (Chapter 6). Notwithstanding the vast profits that can be made, the justifications propose that, however abhorrent this might be, the human health benefits are worth it. These claims have not gone unchallenged, not least because some scientists have pointed to the problems associated with the prediction of human health outcomes from experiments on other animals (e.g. Knight, 2008). Nevertheless, we expect a better and healthier life based on biomedical developments and much of the recent work focuses on genetic research, which is said to herald greater control over human disease, which is leading to increased dependence on experiments using other animals (Brown and Michael, 2001, p. 3). For example, in the UK in 2010 just over 3.7 million new procedures on other animals were started, a rise of 3 per cent on the year before, which 'was largely due to an increase to 1.6 million procedures (+87,000, +6%) in breeding to produce genetically modified (GM) animals and harmful mutants (HM), mainly mice (+77,000)' (Home Office, 2011, p. 7).

The key momentum of biomedical science is the alleviation of human health hazards, hence many would argue that such research is justified (for discussion, see Peggs 2009a). For example, Bernard Williams (2006) maintains that it is reasonable that humans stress their own interests over those of other animals because humans are more important to us. This view is fundamental to biomedical ethics because, as

Jane Welchman observes, in the biomedical community 'it is widely held that partiality to human interests is not only defensible, but also obligatory' (2003, p. 245) and norms and values associated with the intrinsic importance of (biomedical) scientific research are deeply entrenched in Western societies (Gray, 2003, pp. xiii–xiv). However, it cannot be denied that in such experiments other animals are exposed to risks that would be considered unacceptable for humans (Henry and Pulcino, 2009) and would be considered cruel and often criminal if they were practised on other animals who live outside of these laboratories. As we saw in Chapter 4, other animals are legally ill-treated in their billions every year, 'in agriculture and in research laboratories' (Benton, 1998, p. 171). In biomedical experiments other animals are subjected to a range of procedures that include experiments to assess treatments for human diseases, to test the effects of pharmaceuticals, to measure damage from broken bones and to measure the viability of inter-species organ transplants operations. Looking at the UK alone, the most recent figures reveal that in 69 per cent of all procedures (which include those for biomedical and other purposes) no anaesthetic was used (Home Office, 2011, p. 27). In consequence, Beirne argues, the 'criminal law is a major structural and historical mechanism in the consolidation of institutionalized animal abuse' (1999, p. 129) (Chapter 4).

We do not see what goes on in experiments as we are spatially separated from the laboratories in which they take place (Chapter 5). P. Michael Cohn and James V. Parker (2008) argue that, from this, we should not conclude that there is a 'dirty little secret' about what goes on; other animals used in experiments, they argue, are treated with care and compassion. To be sure, Mike Michael and Lynda Birke note that in their interviews with scientists '...it seemed important to our respondents to convey a belief that British animal experimentation is part of a tradition that has directed much effort and many resources to the care of animals' (1994, p. 195). However, several high-profile cases, such as the unveiling of the breaches of the law in the pig to primate organ transplants conducted by *Imutran* which resulted in illegal cruelty and suffering of other animals (Lyons, 2011), have led even those who agree with biomedical experiments to question what goes on in experimental laboratories. Of course, it is entirely reasonable that humans should want to try to reduce, and hopefully eradicate, the health risks that afflict our lives, but this focus on our own needs limits our moral range to one that is increasingly individualized (Peggs, 2011a, p. 51). Individualization is an important idea in this regard as it refers to the ways in which 'I' (in this case the human) focus on 'me' and 'my' yearnings (Beck and Beck-Gernsheim, 1995, p. 95) and 'my' incessant wants and needs include 'my' wish to diminish or eradicate risks to 'my' health. For Bauman, this 'me'-centred society is characterized by a narrow moral scope in which individuals spend most of their lives thinking about which goals to pursue rather than thinking about the means of achieving those goals (2000, p. 2). However, this is not true of all. There are many groups who are concerned about the wider implications and these come in the form of pressure groups and new social movements devoted to a range of, in this case, causes regarding the rights and welfare of humans and the rights and welfare of other animals.

The moral standing of other animals

I spent some time in Chapter 4 examining Garner's (2005a) continuum of recognition of the moral standing of other animals. I want to explore this in further detail here as it is important when considering the positions that activist groups might take. As we saw in Chapter 4, at one end of the continuum is the view that other animals have no moral status. Garner refers to this position as the 'indirect duty' view, in which other animals are viewed as having no intrinsic value and, as a result, 'their protection is totally dependent on whether it serves human interests to do so' (2005b, p. 119). The central position Garner calls the 'animal welfare' position. This position conforms to the moral orthodoxy as it accords inferior moral status to other animals in comparison with humans, but it does give some moral standing to them (Garner, 2005b, p. 16). At the other end of the continuum is the multiple position of 'animal rights/utilitarianism and contractarianism', which sees other animals as having higher moral standing than that accorded by the moral orthodoxy (Garner, 2005b, p. 33). This multiple position is most obviously associated with the advocacy of other animals, but it is important to draw out the central differences among movements that occupy this position.

Philosophical positions on animal advocacy: Rights

Those who argue in favour of rights build 'protective fences around individuals whose fundamental interests cannot, under normal circumstances, be sacrificed to promote the general welfare' (Garner, 2005a, pp. 95–6). This general assertion obscures a good deal of complexity associated with the meaning of 'rights'. Garner is helpful here as he distinguishes between legal rights and moral rights (Garner, 2005a, p. 96).

Legal rights are those that exist as a result of statutes, whereas moral rights refer to those 'existing as a moral entitlement whatever the prevailing legal system permits and prohibits' (Garner, 2005a, p. 96). Moral rights are subdivided into those that are 'acquired' and those that are 'unacquired'. Acquired rights are not common to all as they exist as the result of a specific agreement between individuals, whereas unacquired rights are universal as they are held 'by whole groups of beings, usually as a result of some general characteristic possessed by all of them' and thus are possessed independent of any specific agreement (Garner, 2005a, p. 96). Garner provides a useful example. The notion of 'human rights' is an important concept in current discourses, and when 'human rights' (and, he adds, 'animal rights') are referred to it is the unacquired moral version that is being invoked (Garner, 2005a, p. 96). Regan (1997) is among the many animal advocates who take a rights position for other animals. As we saw in Chapter 4, he argues that moral rights should be accorded to other animals because they are subjects of a life, which means they have inherent value and consequently cannot be treated as a means to an end. Thus, thinking about other animal experiments, Regan (1997) argues that the benefits that derive from experiments on other animals are unacceptable because other animals do not belong in laboratories and thus placing them there in the hope that they will produce benefits for others violates their rights. Before we go any further, I will briefly look at two critiques of the notion of rights.

Critique of rights

In order to think about critiques of the notion of 'animal rights' I turn to the views of two sociologists, Ted Benton and Keith Tester. In offering a critique of Regan's account of animal rights, Benton suggests that Regan applies a 'conventional emphasis on their universal characteristic (except that they apply to some animals as well as humans)' (Garner, 2005a, p. 98). Benton, like Garner, is troubled that the basis of 'rights' is often unexplored (Garner, 2005a, p. 97). For Benton, Regan has skated over the problem of the 'diversity of moral dilemmas posed by our relations to animals', which leads him to conclude that it is unlikely that 'the strategy of assigning universal rights of a very abstract kind to them would be a sufficient response' (1993, p. 92). For example, protecting the interests of other animals used in experiments, other animals used for food, other animals kept as 'pets', other animals kept in zoos and other animals who live in the wild generates very different moral dilemmas. For example, Garner observes that 'institutionalized exploitation' is the norm for other animals who are used on farms, in laboratories and in zoos, and the interests associated with these other animals are different from those who live freely (2005a, p. 98). Moreover, Garner proposes that the relationships between the other animals and the humans concerned differ according to context. For instance, the human who keeps a dog as a 'pet' has a very different relationship with the dog than the scientist who experiments on a dog; in the case of the former 'individual responsibility can be more easily assigned', but responsibility is much more difficult to assign in the latter situation as a variety of humans are involved in a range of elements of the process of such experiments (Garner, 2005a, p. 98). Consequently, Garner suggests,

> Might it not be more appropriate, then, to talk about acquired or special rights in the context of domestic pets rather than universally accorded natural rights? Might it not also be appropriate... to jettison rights in the case of animals subject to institutional exploitation in favour of an emphasis on tackling the economic and social factors responsible for this exploitation?
>
> (2005a, p. 98)

In the case of having a valid claim against an infringement of rights, Garner suggests that his solution is especially pertinent as it is especially difficult to attach responsibility in the case of the institutionalized exploitation of other animals. He uses the example of agriculture as he asks, who should we attach responsibility to – consumers, farmers, retailers, or someone else in the chain (2005a, p. 101)? Finally, Benton argues, there is a difference between granting rights and the protection of interests. He notes that 'As both women's rights and black liberation campaigns have experienced, prevailing structures of economic, cultural and political power may continue to obstruct the realization of juridically acknowledged rights' (Benton, 1993, p. 95). Garner observes a clear parallel with the treatment of other animals because, he argues, the protection that other animals might gain though rights can only be realized if there is 'a broad societal consensus' that other animals have sufficient moral status to 'justify protection against exploitation' (2005a, p. 102).

The second critique comes from Keith Tester. He is concerned that demands for rights 'play on the paradox of the human as a different kind of animal' (Tester, 1991, p. 88). He suggests that the problem with arguments for the rights of other animals is that they are contradictory because such arguments are organized around the demand for difference and the demand for similitude (1991, pp. 88–9). The rights position that relies in a demand for difference emerged in the late nineteenth century.

Advocates who took this position tended to continue to consume other animals (i.e. they perhaps continued to eat 'meat', hunt foxes and wear 'leather'), and they agreed with the view that humans are different from other animals in that they are self-conscious and reflexive beings. Thus, this position 'stressed the uniquely human and the properly social' (Tester, 1991, p. 89). In contrast, the demand for similitude campaigners followed 'a radically alternative path' as they emphasized the similarities between humans and other animals and in consequence argued that 'we should extend humanity to them' (Franklin, 1999, p. 178). These campaigners argued that humans should be vegetarian and compassionate. Tester runs through a very thorough and interesting history of the protagonists on each side, but this need not detain us here. Our focus is on Tester's views of the current position, which leads us to look back to the late nineteenth century and the work of Henry Salt (1980 (orig 1892)).

According to Tester, Salt reconciled the 'difference' and 'similitude' sides as follows:

> Salt has pulled the two faces of 'Man' together. He simultaneously upheld the position of the Demand for Similitude that 'Man' is naturally one animal among many, and the view of the Demand for Difference that 'Man' is separate from all other creatures. 'Man' is also different because 'he' is able to be more perfect than any other living creature.
>
> (1991, pp. 153–4)

Thus, 'animal rights pushes humans into the order of the beast, while simultaneously pulling us into the starry heaven of perfection and virtuous civilisation' (Tester, 1991, p. 170), and this is clear in the language of animal rights which insists that humans are free to the point where they should not restrict the freedom of other creatures (Franklin, 1999, p. 180). Salt's work was largely forgotten until the 1970s, when Singer published his book *Animal Liberation* (Franklin, 1999, p. 181). Franklin sees Singer's approach as heralding the embracing of animal rights as part of the political agenda (1999, p. 181), a political agenda that centres on animal advocacy. Although Singer is certainly an advocate of other animals, Garner (2005a), among others, argues that it is inaccurate to suggest that he takes an animal rights position in the strict sense. Singer puts forward his position as being that of a utilitarian. Although it is certainly true that this might make little difference to at least some of the other animals who are abused, as they would be free if they were granted freedom by means of rights or through utility, it is important to explore the theoretical differences before moving on to other considerations.

Philosophical positions on animal advocacy: Utilitarianism

The utilitarian position is most often associated with Peter Singer. In adopting this position Singer is intentionally turning away from the idea of 'animal rights'. As we might expect, the utilitarian position is complex and Garner (2005a) identifies a number of different forms of utilitarianism. However, we will proceed by simplifying matters. Benton (above) differentiates between 'protection of interests' and 'granting of rights' and Singer locates his focus firmly in interests. Singer argues that equivalent human and other animal interests ought to be considered equally so that, all things being equal, the preference of an other animal to avoid pain should be treated equally with that of a human's (Garner, 2005a, p. 89). This is not the same as asserting that humans and other animals have rights because, as a utilitarian, Singer proposes that 'we ought to aggregate preferences so as to arrive at the situation where their satisfaction is maximised' (Garner, 2005a, p. 89). The utilitarian argument centres on pleasure and pain and the position does not view painless killing as intrinsically wrong, especially if wider benefits are likely to follow (Garner, 2005a, p. 90). Rather, Singer argues that these considerations must be applied to humans and other animals equally, so there must be no resorting to speciesism. He emphasizes that the interests of humans and other animals must be considered similarly (Garner, 2005a, p. 95). Thus, some suffering of other animals and humans can be justified in certain circumstances as, Singer argues, 'if one, or even a dozen animals had to suffer in order to save thousands I would think it right and in accordance with equal consideration of interests that they should do so' (1993, p. 58). This position has been criticized on a number of counts. Singer's approach, and the utilitarian position more generally, has been criticized for arguing that the suffering and killing of other animals, and by extension humans, can be justified if, under equal consideration, more interests are fulfilled by the actions undertaken. As Singer puts it, 'the wrong done to the person killed is merely one factor to be taken into account, and, the preference of the victim could sometimes be outweighed by the preference of others' (1993, p. 95). Linzey (2009) is outraged by this, and he speaks for many. Linzey's position is that 'we would be as disinclined to support painful experimentation on animals as we would be disinclined to support the torture of human subjects, no matter how "beneficial" the results might be in either case' (2009, p. 156). So, the utilitarian position on the advocacy of other animals centres on securing the best possible result in any given situation; however, what is best for the many is not necessarily best for the individual. This brings us to contractarianism.

Philosophical positions on animal advocacy: Contractarianism

The contractarian position is most usually associated with John Rawls' theory of justice (Garner, 2005a, p. 86). As we saw in Chapter 4, this position asks which principles of justice would an individual decide upon if they had no knowledge of their own circumstances (Garner, 2005a, p. 83), that is if the individual were shrouded in a 'veil of ignorance' (Rowlands, 2002, p. 59) that masked their gender, ethnic origin, sexuality, class, religion and so forth. Rawls calls this 'the original position' and it is from this position of ignorance that, he argues, 'we should try to work out what an ideal human society should look like' (Rowlands, 2002, p. 59). Mark Rowlands develops this idea with his notion of the 'impartial position' which, he contends, 'is the version of the original position shaped by the principles of equality and desert' (2002, p. 60). He argues that the impartial position encompasses 'equality' because 'If you don't know who you are, then you have no way of being biased towards yourself' (Rowlands, 2002, p. 60). The impartial position also covers 'desert' because this refers to the features over which one has no control, such as ethnic origin and gender (Rowlands, 2002, p. 60). Consequently, the veil of ignorance that is central to Rowlands' impartial position involves ignorance of those properties 'for which the choice of whether or not we have them is not ours but nature's' (2002, p. 61). Knowledge of one's species would also be masked behind the veil of ignorance. So if a human is invited to imagine that they do not know which species they are a member of and is then asked 'How would I like the world to be?' (Rowlands, 2002, p. 62), their answer will take into consideration the plight of the nonhuman. For Rowlands' this 'provides the most satisfactory theoretical basis for the attribution of moral rights to nonhumans' and is therefore the 'greatest ally' of animal rights (Rowlands, 1998, p. 3 in Garner, 2005a, p. 86). However, Garner argues, this approach is problematic because 'it cannot do the work that pro-animal exponents want it to do' (2005a, p. 86). For example, the water can be muddied, so to speak, as the principle could be applied to nonanimate objects like seas or mountains (Garner, 2005a, p. 86). In order to avoid such muddying, the contract could be limited to sentient beings, which is the position that Singer might take. However, again there are problems as sentience is a difficult concept in itself. The attribution of sentience depends on human definitions and conceptualizations, which makes the dependence on sentience human-centred and problematic. For example, categorizations of sentience exclude a whole host of invertebrate species of other animals (Peggs, 2010). Another problem with contractarianism more generally is, if those who hold the position take the view that only individuals who can understand and choose to take part in an agreement (i.e. a 'social contract') can have moral rights, the position can be used against other animals. For example, the neo-Cartesian Peter Carruthers argues that 'the capacity for rational agency' is required for the individual to be covered by the social contract and, he suggests, this capacity is only evident in humans (Cavalieri, 2001, p. 82).

Animal rights and animal advocacy

Although the above discussion signifies differences in philosophical positions, it is nonetheless clear that those who hold the positions discussed above are concerned with the well-being of animals (whether those animals are human or nonhuman). Although the term 'animal rights' as it is used in the sociological literature (such as by Franklin and Tester) is not as nuanced as that in the more philosophical literature (associated with Garner, Regan, Singer and Rowlands), I think we can differentiate between animal rights (with a small 'r') as it is applied to the catch-all position of 'animal advocacy' and animal Rights (with a capital 'R'), which is centred on the more complex philosophical application associated with moral entitlement. The notion of 'Rights', as it is philosophically used, is important to us here as knowing the motivations of a 'rights' campaign group, as it is commonly used, demonstrates a good deal about the goals of the group. For example, a welfarist group such as Compassion in World Farming (CWF) can be contrasted with the broad 'rights'-based groups because CWF accepts the moral orthodoxy that humans have a higher moral standing than other animals. As a result, in the view of CWF, other animals can be eaten and farmed; however, they should be treated with compassion. Although Regan, Rowlands and Singer adopt different positions they have in common a fundamental concern with the plight of other animals as they argue, to quote Rowlands, 'our current treatment of animals is very wrong indeed' (2002, p. 177). Thus they all call for action. Such action can be taken on a personal level, for example by becoming a vegetarian or vegan and by not wearing the skins of other animals, but it can also take the form of involvement in groups that campaign for the liberation of other animals, such as SPEAK, a group that campaigns against the use of other animals in experiments. Thus SPEAK goes further than the CWF position. The next section concentrates on 'rights' campaign groups within the context of sociological understandings of social movements.

Social movements and animal rights activism

As an organized network of collective action around the well-being of other animals, the animal rights movement is emblematic of a social movement as it is understood in sociology. The movement is goal-oriented, in that it seeks to secure the rights (in the commonly used sense of the word) of other animals. The movement uses various forms of campaigns (e.g. demonstrations, lobbying members of parliament and letters to the media) in order to achieve this goal. The movement does not function with one voice, rather the 'movement' is comprised of many groups (Maurer, 2002). For example, the movement includes global groups such as People for the Ethical Treatment of Animals (PETA) and the Animal Liberation Front (ALF), national groups such as the UK Royal Society for the Protection of Animals (RSPCA) and Korea Animal Rights Advocates (KARA) and local groups such as Oxford-based SPEAK in the UK and New York City Animal Rights in the USA. However, local groups often have national and international wings, and national groups, like the British Union for the Abolition of Vivisection (BUAV), lead global campaigns. Groups under the umbrella might disagree with each other's methods. For example, the RSPCA has distanced itself from the methods used by the ALF (Baker, 1993).

Social movements exist in a context that includes other movements and counter-movements (Maurer, 2002, p. 47), and the animal rights movement exists alongside other movements that support its goals and others that do not. Social movements associated with animal rights exist to encourage social change (e.g. SPEAK and BUAV), whereas counter-movements to the animal rights perspective exist to resist social change (e.g. Pro-Test, an Oxford-based group that campaigned for experiments on other animals) or to reverse social change (e.g. The Countryside Alliance that seeks to reverse the UK ban on hunting with dogs). Sociologists often classify such social movements as 'new' because they are viewed as different from 'old' political movements that existed to change the economic structure in society (Maurer, 2002, p. 48). Although changes to the economic structure might be part of the aims of some groups, the identifying goal of the animal rights movement is the rights of other animals. The movement has emerged in common with other social movements, because standard political engagement (e.g. via elections) is sporadic and offers basic choices that do not centre on the goal of the rights of other animals. They often dissociate themselves from what Beck (1992) calls 'the old collective actors' such as trade unions and political parties, because new forms of reflexivity have enabled them to get involved and because they see the old collective actors as ignoring their interests (Beck, 1992). Thus social movements like the animal rights movement work at the level of subpolitics and their new reflexivity is grounded in humans as reflexive actors who want to change the ways things are. For Randall Collins, Durkheim's analysis of religious forms is pertinent here as at the centre of every social movement is what Durkheim called 'collective effervescence', which is the energy that emerges from a group of people (2001, p. 28). This energy is the product of a 'high ritual density', which consists in a mutual focus of attention that gives the group a feeling of solidarity and a sense of morality (Collins, 2001, p. 28). Thus, the animal rights movement, like any social movement, feels it can, and indeed does, affect public awareness of issues such as experiments on other animals (e.g. by showing undercover films of such experiments).

The term 'liberation' is often used intermitently with the 'term' rights in, for example, media discourses about movements devoted to the advocacy of other animals. In terms of discourses, 'liberation' is a problematic word when used in conjunction with 'animal', not least because of the negative connotations that have been associated with the word through such media discourses. Audiences not involved in animal rights gain much of their understanding of the movement from news media reports. This is not unusual. Many, if not most, of our experiences are mediated; it is through mediation that our views are shaped because it is through mediation that we learn about society (Giddens, 1991, p. 187). In Chapter 4 I referred to analyses of the role of the news media in spearheading moral panics; there I centred on dogs who are labelled 'dangerous dogs'. In respect of animal rights Baker observes what he calls a 'growing hysteria' in places like the media about the dangers of 'animal rights activism' (1993, p. 196). For example, he refers to an article published in 1986 in the UK newspaper *The Star*, which was headlined '21 things you never knew about the animal fanatics' and that was accompanied by the picture of a human head concealed in a balaclava (Baker, 1993, p. 196). This and further news media articles embraced the image of the faceless 'fanatic', and other articles were accompanied by headlines like Terror Alert, which made connections between the ALF and the Irish Republican Army (IRA) (Baker, 1993, p. 196). More recently, in 2000, the BBC ran an article entitled 'Animal Rights: Terror Tactics' in which it was argued that 'Animal rights extremists launched more than 1200 attacks last year – terrorising their victims and causing £2.6m of damage to property' (BBC News, 2000). It seems that the balaclava-clad image of the animal rights 'terrorist' has stuck. Paradoxically, observes Baker, the balaclava-clad image was one to which elements of the animal rights movement (most notably the ALF) assented, but not as that of a terrorist rather as a 'heroic Lone Ranger' (1993, p. 196). Certainly, the more recent ecological agenda has resulted in elements of the news media wishing to portray a concern for the environment, associated with a compassion for other animals and worries about abuse and, most usually, species extinction. Nevertheless, the message is usually 'On the one side thugs and fanatics; on the other side everyone else' (Baker, 1993, p. 206).

Common perceptions of animal rights activism overlook the range of activisms that might be covered by the broad umbrella of 'the movement'. Rowlands assembles this range of activisms into three main categories: lifestyle changes (e.g. refusing to buy products tested on other animals), spreading the word (e.g. demonstrating outside laboratories in which experiments on other animals take place) and civil disobedience (e.g. breaking into a laboratory and destroying equipment used in experiments on other animals) (2002, pp. 177–8). Such activities have been the mainstay of successive animal rights campaigns, and for the activists their chosen activity is appropriate because 'animal rights appeared as a natural truth, something that exists out there' (Franklin, 1999, p. 181). However, Tester argues that animal rights activists have failed to see that animal rights 'are made and not found' (1991, p. 195). For Tester animal rights is an idea, a social construction. Tester's analysis is contentious in a number of ways. One oft-quoted problem with his ideas is the presumptions he makes about the wishes of activists to distance themselves from other animals (by not eating 'meat' and by not wearing 'leather', for example) along with the implicit suggestion that the giving of rights to other animals shows how different we are and how properly human we are (for a critique of Tester, see Baker, 1993). These points are important and deserve considerable contemplation; however, there is no space to do justice to the debate here. What is valuable here is that Tester's critique of animal rights activists and Garner's multifaceted analysis of Rights point to the ways in which the notion of 'natural rights' is at least problematical. In the next section I briefly explore how I have examined, in my other published work, counterarguments against animal rights as they are associated with experiments on other animals.

Identity politics and experiments on other animals

In Chapter 3 I explored sociological understandings of identity, which centre on similarity and difference. As we saw, identification works within notions of those who we see as similar to us and those we feel are different and 'other' than us. An oppressed group, like other animals, is designated 'other' in terms of differences and in terms of exclusion, for example, from certain moral considerations. Derrida adds that the formation of identity is not just a matter of difference and exclusion, it is exclusion established in *hierarchical* notions of difference fundamental to binary oppositions (such as 'human'/'animal'), in which the first category is defined as superior to the excluded, subordinate, second category (Laclau, 1990, p. 33). So, thinking about the identities 'human' and 'animal', the category 'animal' is defined as subordinate in society (and in sociology), and, as we saw in Chapter 2, this is established in ideas about the 'natural' biological distinctions between 'us' and 'them'. For sure, organic differences exist among *all* animals, but assumptions about a solid human identity are founded in the notion that specific organic differences are most significant, and these are the ones that underpin the binary hierarchical classification of all animals into 'human' and 'animal'. This can lead to political antagonisms that are played out by powerful interests and social movements. In Moya Lloyd's words, that which is presented as 'natural (or constative) [is] something that is a political effect (a performative) secured upon the field of power relations' (2005, p. 39). Lloyd's example is instructive. She observes that 'the identity "woman" becomes political, that is it becomes feminist, when an antagonism emerges with men or patriarchy' (Lloyd, 2005, p. 20). Following this example, what I have argued elsewhere (Peggss, 2009a, p. 89) is that 'the identity "human" becomes a political identity (human primacy)' when there is antagonism with the interests of other animals in a society in which other animals are oppressed. My argument draws in identity politics. Identity politics is an important feature of many social movement campaigns because the goal in identity politics is usually 'the call to "respect" traditionally disadvantaged social groups' (Fuller, 2006, pp. 111–12). Thus, social movements associated with women's liberation oppose the oppression of women and demand the end to that oppression. I argue that there are important differences between identity politics and what I have called *primacy* identity politics (2009a, p. 86). In contrast to the goal of the ending of oppression as associated with identity politics, the goal in *primacy* identity politics is, I argue, the preservation or extension of existing power relations through the continued subjugation of disadvantaged groups (2009a, p. 86). In order to explore this point I centred on Pro-Test, an Oxford-based group that counters animal rights positions. Although in February 2010 Pro-Test wound up its UK operations (Pro-Test, 2006), it is a useful group to look at in regard to its position. The group protests in favour of experiments on other animals in the situation where humans benefit from such experiments. One of the stated aims of Pro-Test is to '... to counter the irrational arguments of anti-vivisectionists by raising public awareness of the benefits of animal research' (Pro-Test, 2006). With identity politics it is the shared experience of and identity of oppression that politicizes those involved and which enables political demands to be made (Lloyd, 2005, pp. 55–6). However, Pro-Test is not formed of oppressed beings; it is formed of humans whose aim is to maintain the oppression of other animals through the continued use of experiments. However laudable the wish might be to aid the welfare of humans, it cannot be denied that such a position is speciesist, it justifies the abuse of other animals and, I argue, it relies in the shared experience of an identity of human primacy. Thus, I contend 'without a shared experience of and acceptance of "supremacy" – a primacy identity – Pro-Test could not begin to articulate political demands based in the acceptable exploitation and death of nonhuman animals' (Peggs, 2009a, p. 91). Accordingly, for Pro-Test (2006), experiments on other animals are justifiable on the grounds that they are 'crucially necessary' for the 'welfare of humankind'. Without a doubt, humans have interests but other animals have interests as well. As conscious beings, other animals have desires and preferences but they are treated as 'acceptable losses' (Rowlands, 2002) because '[e]ven if we think that animals can suffer pain, and even death, but humans can suffer more, in the event of a conflict of interests, we would be justified in choosing to sacrifice the interests of animals' (Garner, 2005a, p. 23). Pro-Test's view seems to be located on the central position of Garner's continuum; that is the 'animal welfare' position, which assigns inferior moral status to other animals in comparison with humans, but it does give some moral standing to them (Garner, 2005ap. 16). However, I think Pro-Test's position is a form of what I will call 'weak' moral orthodoxy as Pro-Test's position is less 'animal welfarist' than, for example, that taken by CWF.

Concluding remarks: Animal rights and deconstruction

As we have seen, Garner suggests that there are problems with the 'conventional emphasis' on the 'universal characteristic' of rights as the basis of 'Rights' is often unexplored (Garner, 2005a, p. 97). Garner has used one approach for exploring Rights, another approach might be taken by deconstructionists. A deconstructionist approach to R/rights carries with it its own difficulties; there are considerable problems associated with combining a deconstructionist analysis with a Rights or a rights position against experiments on other animals (Peggs, 2009a). The deconstructionist approach is especially associated with Derrida, whose work I have returned to a number of times in this book. Derrida (1982) argues that we do not have direct access to reality, and thus we must assume there is no reality to the Rights or rights of other animals or of humans. Deconstructionist accounts are associated with liberation from moral standards (Bauman, 1993), not least because deconstructionists argue that since 'ethics is contextual' there are no firm grounds for any particular ethical belief (Crawford, 1998, p. 121). However, as I have suggested elsewhere (Peggs, 2009a), Bauman argues against an 'overtly relativistic and in the end nihilistic view of morality' (1993, p. 12), because for him deconstruction enables us to reflect on that which we take for granted. What we might say is that it encourages us to use our sociological imagination. Thus, it can help with the exposure of the notion of the 'the other' as being inferior and, in consequence, can challenge the idea that other animals can be justly used in experiments. This is because a deconstructionist or

> postmodern ethics would be one that readmits the Other as a neighbour...back from the wasteland of calculated interests to which it had been exiled...an ethics that recasts the Other as the crucial character in the process through which the moral self comes into its own.
> (Bauman, 1993, p. 84)

This is a good starting point for thinking about experiments on other animals and the ways in which sociology might challenge the exploitation of other animals, the subject of the final chapter.

9
Conclusions: Sociology for Other Animals

Introduction

As we have seen, societies are broader than the human. Back in the 1970s, Bryant (1979) argued that other animals play important roles in society and, consequently, are relevant for sociological study. His zoological connection revealed a range of ways in which other animals are central to human societies, and the preceding chapters have shown that this is indeed the case. Other animals are everywhere. Other animals are central to every aspect of our lives: in what we say and what we do; in what we eat and what we wear; in our industry and in our leisure time; in what we worship and what we despise; in how we progress and how we do not. Other animals are central to who we think we are and who we think we are not. Moreover, we are central to the lives of other animals, most notably in our oppression of them. We take away their selves, take away their space, take away their freedom and take away their lives; yet still they seem invisible to us. Sociology is part of the seeing of other animals and, although still somewhat marginal to sociology, that sociological seeing is changing things in sociology. However, it is not just seeing other animals that should be a central component of sociology; the role of sociology in countering the oppression of other animals is also pertinent. This returns me to Bryant.

Clifton Bryant: The zoological 'correction'

As we have seen, Bryant notes that sociologists 'have tended not to recognize, to overlook, to ignore, or to neglect ... the influence of animals, or their import for, our social behavior, our relationships with other humans, and the directions which our social enterprise often takes' (1979, p. 399). He challenged this by arguing that we, as sociologists, should take into account the 'zoological connection' in society. In this regard he was especially interested in the ways in which other animals influence humans in society, from the 'zoological flavor' of human language to the position of other animals in human culture. This is all well and good, but if we peruse his article in a little more detail it becomes evident that Bryant's zoological connection largely accepts the abuse of other animals; in that sense he seems to skate over a central relation that humans have with other animals and in doing so other animals are notably absent. This is abundantly clear in his discussion of human leisure pursuits. In his listing of zoologically connected human pastimes he notes:

> Among our favorite recreational pursuits are horse and dog racing, cock-fighting, hunting and fishing, bird-watching, horseback riding, attending dog and other animal shows, visiting zoos, attending circuses and rodeos, tropical fish raising, practicing taxidermy, and leering at Playboy bunnies.
>
> (Bryant, 1979, p. 403)

There is no critical discussion of these pursuits that displays an ethical stance to humans hunting fish or other animals, watching cocks fight to the death, or constructing other animals as spectacles to be gazed at in zoos, circuses and dog shows. Bryant seems to overlook the oppression of other animals (and the oppression of women who work as 'Playboy bunnies' for that matter). Similarly, in his discussion of the contamination of oyster beds in Chesapeake Bay he is concerned about the effects of this contamination on human consumption and how this might put 'large numbers of watermen out of work' (1979, p. 405). It is dreadful that people should lose their jobs, but Bryant does not seem to be concerned at all about the oysters who have been contaminated. This anthropocentric view is reflected in his analysis of legislation designed to aid the welfare of other animals, and the movements that centre their campaigns in animal rights. He says:

> It may be recalled that concern for whales in our society, and activism in this regard, resulted in legislation that caused the demise of an entire industry, the American whaling industry. Similar concern for dolphins is having its impact on the tuna fishing industry

because of resultant legislation that severely handicaps tuna fishing techniques.

(Bryant, 1979, p. 406)

He does not seem to be concerned about the plight of the whales, dolphins or tunas who die as a result of these industries. Although he mentions these other animals, their lives and their suffering are overlooked in the analysis, in consequence they are invisible. He seems to be especially perturbed by the achievements of animal rights movements. In this regard he laments that:

In Europe 'animal liberation' already has achieved considerable success and animal husbandry there is burdened by an incredible array of regulatory legislation motivated by animal welfare concerns.

(1979, pp. 406–7)

Bryant seems to be only concerned about the 'burdens' that have been put on humans; the other animals who live and die in the most appalling of circumstances do not seem to be an issue. He seems most scathing about the effects of 'animal liberation' on using other animals in experiments. For example, he comments that 'The monkey ban is creating serious concern among many research scientists who fear that some research programs may be imperilled' (1979, p. 407). No mention of the imperilled monkeys or other animals who suffer and die during these research programmes! I am not really knocking Bryant; his article is a product of its time, and indeed was ahead of its time. I agree with Bryant, other animals are certainly an appropriate subject matter for sociology, but the role for sociology in 'animals in society' does not have to stop there. In addition to the zoological connection, for me, sociology could think about a zoological 'correction'. This brings me to a consideration of the purpose of sociology, something that is of great interest to me and which I have discussed elsewhere (Peggs, 2011b).

What is the purpose of sociology?

In this book I have centred on the position of other animals in society and their position in sociology. What I have not done thus far is discuss the purpose of sociology, which I consider to be highly relevant here. I have considered this elsewhere (Peggs, 2011b), and in order to think about the purpose of sociology in this work I explored the opposing viewpoints of Alvin Gouldner (1975), Martyn Hammersley (1998) and Michael Burawoy (2005). Hammersley's position is a good starting point because he puts forward a distinction between what he sees as two related purposes of sociology: the functional purpose of sociology (what purpose *does* it serve to the world) and the moral purpose of sociology (what purpose *should* it serve) (1998, p. 1.3). In setting out his dual purposes of sociology Hammersley critiques Gouldner's (1975) reflexive sociological project. Hammersley (1998) commends Gouldner (1975) for his suggestion that sociology should reflect upon itself and on the social contexts of which it is a part. Gouldner's (1975) reflections led him to conclude that sociology has two purposes: a functional purpose and a moral purpose but, Hammersley (1998) argues, Gouldner does not sufficiently distinguish between these two purposes. Hammersley wants more clarity about the differences between these two purposes, and he sets about trying to do this himself. He explains that the functional purpose of sociology is to be 'no more than a source of specialised factual knowledge about the world' (Hammersley, 1998, abstract). Hammersley acknowledges his debt to Comte who, as we saw in Chapter 1, described sociology as the scientific study of society as *it really is*. This seemingly simple proposition leads to a dilemma because, even if we accept the idea that sociology should centre on the factual (that is what *is* going on in society) first we must decide which conceptualization of society we are talking about (Peggs, 2011b). Sociologists differ. Mead thought that society is inhabited by humans alone whereas Bryant observes that societies are populated by other animals as well. I agree with Bryant on this. However, it seems to me that Bryant might argue that he is taking a purely factual approach in his notion of a zoological connection, where he describes the seemingly endless array of ways in which other animals enter our lives. Is this 'factual' approach the only approach that sociology should take? Gouldner criticizes the factual-only purpose for sociology as he does not see sociology as a scientific and objective enterprise that can disregard questions of morality (1975, p. 25). He argues that the purpose of sociology is that of providing a basis for 'right living' (Gouldner, 1970), which is a manifestation of his idea of the moral purpose of sociology. Here, Gouldner is advocating that sociology should be able to tell us what is right and wrong with the world and with the societies in which we live. Thus, thinking about this book, sociology should do more than describe the relations between humans and other animals; it should have something to say about the rights and the wrongs of these relations and how sociology might be used to change things for the better. However, Hammersley disagrees with this 'grand conception' of sociology's role, he argues for a more 'modest' role (1998, p. 3.1). He is adamant that sociologists should focus exclusively on the factual, and we should 'do this in a way that takes no account of whether we believe what we are studying is good or bad' (1998, p. 4.5). Put simply, for Hammersley, 'Sociology cannot tell us what is wrong and what should be done' (1998, p. 4.1).

This brings me to Burawoy. Like Gouldner, Burawoy (2005) sees room for making a 'better world' through sociology (2005, p. 25). Burawoy's sociological project lies in his promotion of a sociology that embraces four related spheres: professional sociology (i.e. based on accumulated bodies of knowledge), policy-oriented sociology (i.e. on service of a goal defined by a client), public sociology (i.e. on conversation with the public) and critical sociology (i.e. which 'attempts to make professional sociology aware of its biases, and silences and promoting new research areas') (2005, p. 11). He observes that classic sociological studies have sought to make a better world (Burawoy, 2005, p. 6) and points to the call from Dorothy Smith (1987a) for 'A Sociology for Women'. In this classic sociological work, Smith criticizes sociology for taking the male standpoint. Smith's (1987a) mission is to generate a sociology that will reclaim the voice of women and all other disenfranchised members of society. Accordingly, she advocates a woman's standpoint that 'discredits sociology's claim to objective knowledge'. Beth Humphries (1997) calls this the 'emancipatory potential of sociology' as it is a sociology that sits on the side of the oppressed. This emancipatory potential brings into focus ideas about the place of values and value freedom in sociology.

Other animals, values and the emancipatory potential of sociology

Bryant's (1979) call for the 'zoological connection' provides an insight into 'animal studies' in sociology. 'Animal studies' is a broad field, which covers a range of scholars from a variety of disciplines who engage in both emancipatory and nonemancipatory studies. Much animal studies sociology is of the nonemancipatory type and Arluke argues that 'Like the sociology of medicine, sociological research in animal studies has been driven by our own research agenda rather than by the needs of animal advocates and nonscholars who work directly with or for animals' (2002, p. 2), and by the needs of other animals themselves (Peggs, 2011b). However, as we have seen, some sociologists see problems with the expression of values in sociology. For example, Hammersley argues that sociology has no role to play in telling us what is wrong and what should be done to put it right (1998, p. 4.1). Nevertheless, in areas of sociology such as in feminist theory, there is an obvious moral vision about how to change the world for the better for women and for other oppressed groups (e.g. Smith, 1987a). Sociologists have pointed to the impossibility of objectivity in any area that we study and, in consequence, taking a moral stance might be seen simply as an overt explication of subjectivity. They also might ask whether it is morally defensible to endeavour to be objective about the suffering of any being, whether human or other animal. The previous chapters reveal that humans gain a great deal from the oppression of other animals and thus we have much to lose in their emancipation. In taking a human standpoint, a sociology that tries to be objective about the suffering and abuse of other animals is not being objective at all; it is coming from the position of the oppressor. In her introduction to Smith's (1987b) radical critique of sociology, Sandra Harding argues that Smith's sociology for women 'discredits sociology's claim to constitute an objective knowledge independent of the sociologist's situation' (1987, p. 84). For me, sociology for humans and other animals also does the same discrediting of traditional sociology's claims to objectivity, as sociologists are human. In consequence, rather than being objective, sociological knowledge as we understand it is grounded in humanness (Peggs, 2011b).

Sociology for humans and other animals

Critical sociological work, such as that by Smith (1987a, 1987b), shows that sociological perspectives are partial. Sociology has often ignored other animals. However, other animals are now more visible in sociology than ever before, but sociological accounts do not necessarily address the oppression of other animals. The human standpoint is one that takes for granted the human way of looking at the world (Peggs, 2011b). Although an ambiguous and contested term, posthumanism could point to a way forward. Cary Wolfe sees posthumanist thinking as a challenge to narrow definitions of what it means to be human or nonhuman and, through this challenge, it 'forces us to rethink our taken-for-granted modes of human experience' (2010, p. xxv). The sociologist Erika Cudworth (2011) adopts a posthumanist approach in her work. She centres her analysis on the 'matrix of domination' that specifies relations between humans and other animals (2011, p. 14), and thus makes clear that 'species must be included in the agenda for critical social science' (2011, p. 13). The chapters in this book show that other animals are exploited and oppressed by humans, and humans benefit from the oppression and exploitation of other animals. This system of domination, Adams (1995) argues, interconnects with other systems of domination. How can sociologists introduce balance and objectivity into explorations of such oppressions? Hammersley (1998) advises that sociologists should describe the facts. But, the 'facts' we describe are described from a particular standpoint. Standpoint sociologists recognize that this is the case.

David Nibert notes that 'Oppression is usually naturalized – that is, it is made to appear as a normal and innate part of world existence' (2003, p. 18). Sociology more often than not seeks to question notions of the naturalness of the oppression of 'devalued groups of humans' (Nibert, 2002, p. 6), but sociology seems to accept as natural (or render invisible) the oppression of other animals. It is surely beyond dispute that other animals are oppressed by humans so, is the human treatment of other animals acceptable for the attention in sociology? Bryant (1979) argues that sociologists should engage with the 'zoological connection'. An aspect of this could be enquiry into the human oppression of other animals. This means that, as sociologists, we would need to confront our own speciesism. Nibert argues as follows.

> Members of the discipline [sociology], who like most other humans in society partake in the privileges derived from entangled oppressions – such as eating and drinking substances derived from the bodies of 'others', wearing their skin and hair, and enjoying the entertainment value their exploitation provides – can do so only by accepting the self-interested realities crafted by powerful agribusiness, pharmaceutical and other industries that rely on public acquiescence in oppressive social arrangements. Privilege is not so easy to give up. Silence, denial and substantial intellectual acrobatics are necessary for oppression in all forms to continue.
> (2003, pp. 20–1)

In this book we have seen that

- sociology can make a significant contribution to our understanding of human relations with other animals,
- sociology can make a significant contribution to our understanding of relations among other animals, and
- the study of other animals in society can make a significant contribution to sociology.

In doing so, we sociologists can confront our own human-centredness. Thus, coming back to the Sanger's (1996) story of the man who smuggled donkeys (Chapter 1), we will not only be unlike the border guard because we will see the donkeys who are being smuggled, but also will be able to see that (as owned, worked and enslaved other animals who are held captive and are treated as consumables who can be bought and sold by humans) donkeys are abused and oppressed, and, importantly, we can use sociology to try to challenge this.

References

Adams, C. J. (1990). *The Politics of Meat.* Cambridge: Polity.
Adams, C. J. (1995). *Neither Man nor Beast: Feminism and the Defense of Animals.* New York: Continuum.
Alexander, V. D. (2003). *Sociology of the Arts: Exploring Fine and Popular Forms.* Oxford: Blackwell.
Alger, J. M. and Alger, S. F. (1997). Beyond Mead: Symbolic Interaction between Humans and Felines. *Society and Animals* 5 (*1*). 65–81.
Alger, J. M. and Alger, S. F. (2003). Drawing the Line. *International Journal of Sociology and Social Policy* 23 (*3*). 69–93.
Anderson, K. (1998a). Animal Domestication in Geographic Perspective. *Society and Animals* 6 (*2*). 119–35.
Anderson, K. (1998b). Animals, Science, and Spectacle in the City. In J. A. Wolch and J. Emel (Eds), *Animal Geographies: Place, Politics, and Identity in the Nature-Culture Borderlands* (pp. 27–50). London: Verso.
Anderson, P. K. (2003). A Bird in the House: An Anthropological Perspective on Companion Parrots. *Society & Animals* 11 (*4*). 393–418.
Arluke, A. (2002). A Sociology of Sociological Animal Studies. *Society and Animals* 10 (*4*). 369–74.
Arluke, A. and Hafferty, F. (1996). From Apprehension to Fascination with Dog Lab: The Uses of Absolutions by Medical Students. *Journal of Contemporary Ethnography* 25. 201–25.
Arluke, A. and Sanders, C. R. (1996). *Regarding Animals.* Philadelphia: Temple University.
Armstrong, S. J. and Botzler, R. G. (1993). *Environmental Ethics: Divergence and Convergence.* New York: McGraw-Hill.
Ascione, F. R. (1998). Battered Women's Reports of their Partners and their Children's Cruelty to Animals. *Journal of Emotional Abuse* 1. 119–33.
Ascione, F. R. (2008). Animal Abuse: The Continuing Evolution of Theory, Research and Application. In F. R. Ascione (Ed.), *The International Handbook on Animal Abuse and Cruelty: Theory, Research and Application* (pp. 473–8). Indiana: Purdue University Press.
Ascione, F. R., Weber, C. V. and Wood, D. S. (1997). The Abuse of Animals and Domestic Violence: A National Survey of Shelters for Women Who are Battered. *Society and Animals* 5 (*3*). 205–18.
Baker, S. (1993). *Picturing the Beast: Animals, Identity and Representation.* Champaign: University of Illinois Press.
Barthes, R. (1993). *Mythologies.* London: Vintage.
Bauman, Z. (1993). *Postmodern Ethics.* Oxford: Blackwell.
Bauman, Z. (2000). *Liquid Modernity.* Cambridge: Polity.
Bauman, Z. (2006). *Liquid Fear.* Cambridge: Polity.
BBC News. (2000). *Animal Rights, Terror Tactics.* 30 August 2000. Retrieved 13 July 2011, from BBC News UK: http://news.bbc.co.uk/1/hi/uk/902751.stm.

Beck, U. (1992). *Risk Society: Towards a New Modernity*. London: SAGE.

Beck, U. and Beck-Gernsheim, E. (1995). *The Normal Chaos of Love*. Cambridge: Polity.

Beirne, P. (1999). For a Nonspeceisist Criminology: Animal Abuse as an Object of Study. *Criminology 37 (1)*. 117–47.

Beirne, P. (2002). Criminology and Animal Studies: A Sociological View. *Society and Animals 10 (4)*. 381–6.

Beirne, P. (2009). *Controlling Animal Abuse: Law, Criminology and Human-Animal Relationships*. Lanham, MD: Rowman and Littlefield.

Beirne, P. and South, N. (Eds) (2007). *Issues in Green Criminology: Confronting Harms against Environments, Humanity and Other Animals*. Devon: Wilan.

Bekoff, M. (2002). *Minding Animals: Awareness, Emotion and Heart*. Oxford: Oxford University Press.

Bentham, J. (1970 (orig 1789)). *An Introduction to the Principles of Morals and Legislation* (Ed. J. H. Burns and H. L. A. Hart). London: Athlone Press.

Benton, T. (1993). *Natural Relations, Ecology, Social Justice and Animal Rights*. London: Verso.

Benton, T. (1998). Rights and Justice on a Shared Planet: More Rights or New Relations? *Theoretical Criminology 2 (2)*. 149–75.

Berger, J. (2009). *Why Look at Animals*. London: Penguin.

Berger, P. (1963). *An Invitation to Sociology*. New York: Anchor Books.

Berger, P. and Luckmann, T. (1967). *The Social Construction of Reality: A Treatise in the Sociology of Knowledge*. Garden City, NY: Anchor Books.

Bergman, C. (1990). *Wild Echoes: Encounters with the Most Endangered Animals in North America*. Toronto: McGraw-Hill.

Bierstedt, R. (1960). Sociology and Humane Learning. *American Sociological Review 25 (1)*. 3–9.

Birke, L., Arluke, A. and Michael, M. (2007). *The Sacrifice: How Scientific Experiments Transform Animals and People*. Indiana: Purdue University Press.

Blaney, D. P. (2002). *The Changing Landscape of U.S. Milk Production. Statistical Bulletin Number 978*. Washington: U.S. Department of Agriculture.

Blumer, H. (1969). *Symbolic Interactionism: Perspective and Method*. Englewood Cliffs, NJ: Prentice-Hall.

Bokonyi, S. (1969). Archaeology: Problems and Methods of Recognizing Animal Domestication. In P. Ucko and G. Dimbleby (Eds), *The Domestication and Exploitation of Plants and Animals* (pp. 219–29). London: Duckworth.

Bourdieu, P. (1984). *Distinction: A Social Critique of the Judgement of Taste*. Cambridge, MA: Harvard University Press.

British Hen Welfare Trust. (2010). *Home Page*. Retrieved 4 July 2011, from British Hen Welfare Trust: http://www.bhwt.org.uk/

Brown, N. and Michael. M. (2001). Switching Between Science and Culture in Transpecies Transplantation. *Science, Technology and Human Values. 1 (Winter 2001)*. 3–22.

Bryant, C. D. (1979). The Zoological Connection: Animal-Related Human Behavior. *Social Forces 58 (2)*. 399–421.

Buller, H. and Morris, C. (2007). Animals and Society. In J. N. Pretty, A. Ball, T. Benton, J. Guivant, D. R. Lee, D. Orr, M. Pfeffer and H. Ward (Eds), *The SAGE Handbook of Environment and Society* (pp. 471–84). London: SAGE.

References

Burawoy, M. (2005). For Public Sociology. *American Sociological Review* 70 (1). 4–28.
Burridge, J. (2008). 'Hunting Is Not Just for Blood-Thirsty Toffs': The Countryside Alliance and the Visual Rhetoric of a Poster Campaign. *Text and Talk* 28 (1). 31–53.
Busfield, J. (2006). Pills, Power, People: Sociological Understandings of the Pharmaceutical Industry. *Sociology* 40 (2). 297–314.
Butler, J. (1999). *Gender Trouble*. 2nd edn. London: Routledge.
Calarco, M. (2008). *Zoographies: The Question of the Animal from Heidegger to Derrida*. New York: Columbia University Press.
Calhoun, C. G., Gerteis, J., Moody, J., Pfaff, S., Schmidt, K. and Virk, I. (2002). Introduction. In C. G. Calhoun, J. Gerteis, J. Moody, S. Pfaff, K. Schmidt and I. Virk. (Eds), *Classical Social Theory* (pp. 1–18). Malden, MA: Blackwell.
Cavalieri, P. (2001). *The Animal Question: Why Nonhuman Animals Deserve Human Rights*. Oxford: Oxford University Press.
Cazaux, G. (2007). Labelling Animals: Non-Speciesist Criminology and Techniques to Identify Other Animals. In P. Beirne and N. South (Eds), *Issues in Green Criminology: Confronting Harms against Environments, Humanity and Other Animals* (pp. 87–113). Cullompton, Devon: Willan.
Chambers. (1994). *The Chambers Dictionary*. Edinburgh: Harrap.
Cohen, R. and Kennedy, P. (2000). *Global Sociology*. Hampshire: Macmillan.
Cohen, S. (2003 (orig 1972)). *Folk Devils and Moral Panics: The Creation of Mods and Rockers*. 3rd edn. London: Routledge.
Cohn, P. M. and Parker, J. V. (2008). *The Animal Research War*. Basingstoke: Palgrave Macmillan.
Cole, M. and Morgan, K. (2011). Vegaphobia: Derogatory Discourses of Veganism and the Reproduction of Speciesism in UK National Newspapers. *The British Journal of Sociology*, 134–53.
Collins, R. (2001). Social Movements and the Focus of Emotional Attention. In J. Goodwin, J. M. Jasper and F. J. Polletta (Eds), *Passionate Politics: Emotions and Social Movements* (pp. 27–44). Chicago: University of Chicago Press.
Commission of the European Communities. (2007). *Fifth Statistical Report on Laboratory Animals*. Brussels: Council of European Communities Publications Office.
Compassion in World Farming. (2008). *Colorado Bans the Veal Crate and the Gestation Crate*. 19 May 2008. Retrieved 30 June 2011, from Compassion in World Farming: http://www.ciwf.org.uk/news/factory_farming/colorado_bans_veal_crates.aspx.
Comte, A. (1975 (orig 1851–54)). *Auguste Comte and Positivism: The Essential Writings* (Ed. G. Lenzer). New York: Harper Torchbooks.
Connell, A. (2011). *Behind the Scenes at Linda McCartney*. Retrieved 4 July 2011, from Linda McCartney Foods: http://www.lindamccartneyfoods.co.uk/_Assets/PDFs/VegSocVisitFactory.pdf.
Council of the European Communities. (1986). *Council Directive 86/609/EEC of 24 November 1986 on the Approximation of Laws, Regulations and Administrative Provisions of the Member States Regarding the Protection of Animals Used for Experimental and Other Scientific Purposes*. Brussels: Council of European Communities Publications Office.

Craig, I. W. (2007). The Importance of Stress and Genetic Variation in Human Aggression. *BioEssays 29 (3)*. 227–36.

Crawford, N. C. (1998). Postmodern Ethical Conditions and a Critical Response. *Ethics and International Affairs 12 (1)*. 121–40.

Cudworth, E. (2011). *Social Lives with Other Animals: Tales of Sex, Death and Love*. Basingstoke: Palgrave Macmillan.

Darnton, R. (1985). *The Great Cat Massacre and Other Episodes in French Cultural History*. Harmondsworth: Penguin.

Darwin, C. (1998). *The Origin of Species*. Oxford: Oxford University Press.

Davis, C. G and Lin, B-L. (2005). *Factors Affecting U.S. Beef Consumption*. Washington: Economic Research Service/United States Department of Agriculture.

Davis, K. and Moore, W.(1945). Some Principles of Stratification. *American Sociological Review 10 (2)*. 242–9.

de Botton, A. (2011). In Praise of Zoos. 8 July 2011. Retrieved 8 July 2011, from BBC Radio 4: http://www.bbc.co.uk/programmes/b012942v#synopsis.

Delgado, C. L. (2003). Rising Consumption of Meat and Milk in Developing Countries has Created a New Food Revolution. *The American Society for Nutritional Sciences Journal of Nutrition. Supplement Animal Source Foods to Improve Micronutrient Nutrition in Developing Countries*. 3907–10.

Derrida, J. (1982). *Margins of Philosophy* (Trans. Alan Bass). Brighton: Harvester.

Derrida, J. (2004). The Animal that Therefore I Am (More to Follow). In P. Atterton amd M. Calarco (Eds), *Animal Philosophy: Essential Readings in Continental Thought* (pp. 113–28). London: Continuum.

Descartes, R. (1993). Animals are Machines. In S. J. Armstrong and R. J. Botzler (Eds), *Environmental Ethics: Divergence and Convergence* (pp. 281–5). New York: McGraw-Hill.

Descartes, R. (2007). From the Letters of 1646 and 1649 (Trans. and Ed. Anthony Kenny). In L. Kalof and A. Fitzgerald (Eds), *The Animals Reader: The Essential Classic and Contemporary Writings* (pp. 59–62). Oxford: Berg.

Digard, J.-P. (1990). *L'Homme et les animaux domestiques: Anthropologie d'une passion* [Man and Domestic Animals: The Anthropology of a Passion]. Paris: Fayard.

Donovan, J. (1990). Animal Rights and Feminist Theory. *Signs: Journal of Women in Culture and Society 15 (2)*. 350–75.

Douglas, M. (1970). *Purity and Danger: An Analysis of Concepts of Pollution and Taboo*. Harmondsworth: Penguin.

Douglass, C. (1997). *Bulls, Bullfighting, and Spanish Identities*. Tuscon: University of Arizona Press.

Ducos, P. (1978). Domestication Defined and Methodological Approaches to its Recognition in Faunal Assemblages. In R. H. Meadow and M. A. Zeder (Eds), *Approaches to Fauna Analysis in the Middle East* (pp. 53–6). Cambridge Peabody Museum: Peabody Museum Bulletins No. 2.

Dunlap, R. E. and Catton, W. R. (1979). Environmental Sociology. *Annual Review of Sociology. 5*. 243–73.

Dupre, J. (2002). *Humans and Other Animals*. Oxford: Oxford University Press.

Dupre, J. (2004). Human Kinds and Biological Kinds: Some Similarities and Differences. *Philosophy of Science 71 (5)*. 892–900.

Durkheim, E. (1947). *The Division of Labor in Society* (Trans. G. Simpson). New York: Free Press.

Durkheim, E. (1964 (orig 1885)). *The Rules of the Sociological Method*. New York: Free Press.
Durkheim, E. (1965 (orig 1915)). *The Elementary Forms of Religious Life*. New York: Free Press.
Durkheim, E. (1966 (orig 1897)). *Suicide*. New York: Free Press.
Durkheim, E. (2005). The Dualisim of Human Nature and its Social Conditions. *Durkheimian Studies 2*, 35–45.
Emel, J., Wilbert, C. and Wolch, J. (2002). Animal Geographies. *Society & Animals. 10 (4)*. 407–12.
Evans, R., Gauthier, D. R. and Forsyth, C. J. (1998). Dogfighting: Symbolic Expression and Validation of Masculinity. *Sex Roles 39 (11/12)*. 825–38.
Faver, C. C. and Strand, E. B. (2003). To Leave or to Stay? Battered Women's Concern for Vulnerable Pets. *Journal of Interpersonal Violence. 18 (12)*. 1367–77.
Featherstone, M. (1991). *Consumer Culture and Postmodernism*. London: SAGE.
Ferguson, C. J. (2009). *Violent Crime: Clinical and Social Implications*. London: SAGE.
Fishman, J. A. (1972). *The Sociology of Language: An Interdisciplinary Social Science Approach to Language in Society*. Cambridge, MA: Newbury House Publishers.
Flynn, C. P. (2000). Why Family Professionals Can No Longer Ignore Violence Towards Animals. *Family Relations 49 (1)*. 87–95.
Flynn, C. P. (2008). A Sociological Analysis of Animal Abuse. In F. R. Ascione (Ed.), *The International Handbook on Animal Abuse and Cruelty: Theory, Research and Application* (pp. 155–74). Indiana: Purdue University Press.
Flynn, C. P. (2011). Examining the Links between Animal Abuse and Human Violence. *Crime, Law, and Social Change* (Special Issue: Animal Abuse and Criminology). 55, 453–68.
Forsyth, C. J. and Evans, R. D. (1998). Dogmen: The Rationalization of Deviance. *Society and Animals 6 (3)*. 203–18.
Foucault, M. (1976). *The Birth of the Clinic*. London: Tavistock.
Foucault, M. (1980). *Michel Foucault Power/Knowledge: Selected Interviews and Other Writings 1972–1977* (Ed. C. Gordon. Trans. C. Gordon, L. Marshall, J. Mepham and K. Soper). London: Harvester Wheatsheaf.
Francione, G. (2008). *Animals as Persons: Essays on the Abolition of Animal Exploitation*. New York: Columbia University Press.
Franklin, A. (1999). *Animals and Modern Culture*. London: SAGE.
Freese, J., Allen Li, J. and Wade, L. D. (2003). The Potential Relevances of Biology to Social Inquiry. *Annual Review of Sociology, 29 (August)*. 233–56.
Frost, D., Smith, C. and Francis, F. (2010). *Market Review of the Organic Red Meat Sector in Wales 2009: Summary and Recommendations*. Cardiff: Organic Centre Wales.
Fuller, S. (2006). *The New Sociolgical Imagination*. London: SAGE.
Garner, R. (1998). *Political Animals: Animal Protection Politics in Britain and the United States*. Basingstoke: Palgrave Macmillan.
Garner, R. (2005a). *Animal Ethics*. Cambridge: Polity.
Garner, R. (2005b). *The Political Theory of Animal Rights*. Manchester: Manchester University Press.
Geertz, C. (2003). Thick Description: Towards an Intepretive Theory of Culture. In C. Jencks (Ed.), *Culture: Critical Concepts in Sociology* (pp. 173–96). London: Routledge.

Giddens, A. (1987). *The Nation-State and Violence.* Cambridge: Polity Press.
Giddens, A. (1991). *Modernity and Self Identity.* Cambridge: Polity.
Giddens, A. and Duneier, M. (2000). *Introduction to Sociology.* 3rd edn. New York: W. W. Norton.
Giddens. A. (2009). *Sociology.* 6th edn. Cambridge: Polity.
Goodall, P. (1995). *High Culture, Popular Culture: The Long Debate.* St Leonards: Allen and Unwin.
Goring, C. (1913). *The English Convict: A Statistical Study.* London: HMSO.
Gould, S. J. (1980). *Ever Since Darwin: Reflections in Natural History.* Harmondsworth: Penguin.
Gouldner, A. W. (1970). *The Coming Crisis of Western Sociology.* New York: Basic Books
Gouldner, A. W. (1975). *For Sociology: Renewal and Critique in Sociology Today.* Harmondsworth: Penguin.
Grant, C. (2006). *The No-Nonsense Guide to Animal Rights.* Oxford: The New Internationalist.
Gray, J. (2003). *Straw Dogs: Thoughts on Humans and Other Animals.* London: Granta.
Guo, G. (2006). The Linking of Sociology and Biology. *Social Forces 85 (1).* 145–9.
Hall, S. (1980). *Culture, Media, Lanugages.* London: Hutchinson.
Hall, S. (1996). Introduction: Who Needs Identity? In S. Hall and P. du Gay (Eds), *Questions of Cultural Identity* (pp. 1–19). London: SAGE.
Hammersley, M. (1998) Sociology, What's It For? A Critique of Gouldner. *Sociological Research Online 4 (4).* No pagination.
Haraway, D. (1991). *Simians, Cyborgs, and Women: The Reinvention of Nature.* London: Free Association Books.
Harding, S. (1987) *Feminism and Methodology.* Indiana: Indiana University Press.
Henry, B. and Pulcino, R. (2009). Individual Difference and Study-Specfic Characteristics Influencing Attitudes About the Use of Animals in Medical Research. *Society and Animals 17 (4).* 305–24.
Hickrod, L. and Schmitt, R. (1982). A Naturalistic Study of Interaction and Frame: The Pet as a 'Family Member'. *Urban Life 11,* 55–77.
Hinchliffe, S. (2007). *Geographies of Nature: Societies, Environments, Ecologies.* London: SAGE.
Holloway, L. (2003). 'What a Thing, then, Is This Cow ...': Positioning Domestic Livestock Animals in the Texts and Practices of Small-Scale 'Self-Sufficiency'. *Society and Animals 11 (2).* 145–65.
Home Office. (2011). *Statistics of Scientific Procedures on Living Animals: Great Britain 2010.* London: The Stationary Office.
Humphries, B. (1997). From Critical Thought to Emancipatory Action: Contradictory Research Goals. *Sociological Research Online. 2 (1).* No pagination.
Ingold, T. (1988). *What is an Animal?* London: Unwin Hyman.
Ingold, T. (1994). Introduction. In T. Ingold (Ed.), *What is an Animal?* (pp. 1–16). London: Routledge.
Irvine, L. (2003). George's Bulldog: What Mead's Canine Companion Could Have Told Him About the Self. *Sociological Origins 3 (1).* 46–49.
Irvine, L. (2004). *If You Tame Me: Understanding Our Connection with Animals.* Philadelphia: Temple University Press.

Irvine, L. (2007). The Question of Animal Selves: Implications for Sociological Knowledge and Practice. *Qualitative Sociology Review 3 (1)*. 5–22.

Jasanoff, S. (2005). *Designs on Nature: Science and Democracy in Europe and the United States*. Princeton: Princeton University Press.

Jenkins, R. (2004). *Social Identity*. 2nd edn. London: Routledge.

Jerolmack, C. (2007). Animal Archaelology: Domestic Pigeons and the Nature-Culture Dialectic. *Qualitative Sociology Review 3 (1)*. 74–95.

Jerolmack, C. (2008). How Pigeons Became Rats: The Cultural-Spatial Logic of Problem Animals. *Social Problems 55 (1)*. 72–94.

Kalof, L. and Fitzgerald, A. (2007). *The Animals Reader: The Essential Classic and Contemporary Writings*. Oxford: Berg.

Kalof, L., Fitzgerald, A. and Baralt, L. (2004). Animals, Women, and Weapons: Blurred Sexual Boundaries in the Discourse of Sport Hunting. *Society and Animals 12 (3)*. 237–51.

Kean, H. (2001). Imagining Rabbits and Squirrels in the English Countryside. *Society and Animals 9 (2)*. 163–75.

Kemp, M. (2007). The Human Animal. *History Today 57 (1)*. 34–41.

Klein, N. (2001). *No Logo: Taking Aim at the Brand Bullies*. London: Harper Collins.

Knight, A. (2008). Systematic Reviews of Animal Experiments Demonstrate Poor Contribution Toward Human Health Care. *Reviews on Recent Clinical Trials 3*. 89–96.

Kruse, C. R. (2002). Social Animals: Animal Studies and Sociology. *Society and Animals 10 (4)*. 375–79.

Laclau, E. (1990). *New Reflections on the Revolution of Our Time*. London: Verso.

Latour, B. (1993). *We Have Never Been Modern*. Cambridge: Harvard University Press.

Latour, B. (2004). *Politics of Nature*. Cambridge: Harvard University Press.

Lawrence, E. A. (1994). Conflicting Ideologies: Views of Animal Rights Advocates and Their Opponents. *Society and Animals 2 (2)*. 175–90.

Leavis, Q. D. (1978). *Fiction and the Reading Public*. London: Chatto and Windus.

Lefebvre, H. (1991). *The Production of Space*. London: Blackwell.

Levi-Strauss, C. (1962). *Totemism* (Trans. R. Needham). London: Merlin Press.

Levi-Strauss, C. (1964). *Mythologiques: le Cru et le Cuit*. Paris: Plon.

Linzey, A. (2009). *Why Animal Suffering Matters: Philosophy, Theology, and Practical Ethics*. Oxford: Oxford University Press.

Lloyd, M. (2005). *Beyond Identity Politics: Feminism, Power and Politics*. London: SAGE.

Lovegrove, R. (2007). *Silent Fields: The Long Decline of a Nation's Wildlife*. Oxford: Oxford University Press.

Low, M. (2006). The Social Construction of Space and Gender. *European Journal of Women's Studies 13 (2)*. 119–33.

Lupton, D. (1996). *Food, the Body and the Self*. London: SAGE.

Lynch, M. E. (1988). Sacrifice and the Transformation of the Animal Body into a Scientific Object: Laboratory Culture and Ritual Practice in the Neurosciences. *Social Studies of Science 18 (2)*. 265–89.

Lyons, D. (2011). Protecting Animals versus the Pursuit of Knowledge: The Evolution of the British Animal Research Policy Process. *Society and Animals 19 (4)*. 356–67.

Macionis, J. and Plummer, K. (2005). *Sociology: A Global Introduction*. 3rd edn. Harlow: Pearson Education.

MacNair, R. M. (2001). McDonald's 'Empirical Look at Becoming Vegan'. *Society and Animals* 9 (1). 63–9.

Maher, J. and Pierpoint, H. (2011). Friends, Status Symbols and Weapons: The Use of Dogs by Youth Groups and Youth Gangs. *Crime, Law and Social Change*. 55 (5). 405–20.

Malamud, R. (1998). *Reading Zoos: Representations of Animals and Captivity*. London: Macmillan.

Marino, L., Lilienfeld, S. O., Malamud, R., Nobis, N. and Broglio, R. (2010). Do Zoos and Aquariums Promote Attitude Change in Visitors? A Critical Evaluation of the American Zoo and Aquarium Study. *Society and Animals* 18 (2). 126–38.

Marvin, G. (1988). *Bullfight*. New York: Basil Blackwell.

Marvin, G. (2001). Cultured Killers: Creating and Representing Foxhounds. *Society & Animals* 9 (3). 273–92.

Marvin, G. (2002). Unspeakability, Inedibility, and Structures of Pursuit in the English Foxhunt. In N. Rothfels (Ed.), *Representing Animals* (pp. 139–58). Bloomington, IN: Indiana University Press.

Marx, K. and Engels, F. (2009 (orig 1848)). *The Communist Manifesto*. Middlesex: The Echo Library.

Marx, L. (1994). The Environment and the Two Cultures Divide. In J. R. Fleming (Ed.), *Science, Technology and the Environment: Multidiscplinary Perspectives* (pp. 3–21). Akron, OH: University of Akron Press.

Mason, J. and Finelli, M. (2006). Brave New Farm? In P. Singer (Ed.), *The Defense of Animals: The Second Wave* (pp. 104–22). Oxford and Malden, MA: Carlton Blackwell.

Massey, D. (2005). *For Space*. London: SAGE.

Maurer, D. (2002). *Vegetarianism: Movement or Moment?* Temple: Philadelphia University Press.

McDonald, B. (2000). Once You Know Something, You Can't Not Know It: An Empirical Look at Becoming Vegan. *Society and Animals* (1). 1–23.

McKibben, B. (2003). *The End of Nature: Humanity, Climate Change and the Natural World*. 2nd edn. London: Bloomsbury.

McSpotlight. (N/D). *The Issues: Animals*. Retrieved 4 July 2011, from McSpotlight: http://www.mcspotlight.org/issues/animals/index.html.

Mead, G. H. (1934). *Mind, Self and Society. From the Standpoint of a Social Behaviorist* (Ed. C. W. Morriss). Chicago: University of Chicago Press.

Mead, G. H. (1964). *George Herbert Mead on Social Psychology* (Ed. Anslem Strauss). Chicago: University of Chicago Press.

Merchant, C. (1982). *The Death of Nature: Ecology, Women and the Scientific Revolution*. London: Wildwood House.

Michael, M. and Birke, L. (1994). Accounting for Animal Experiments: Identity and Disreputable 'Others'. *Science, Technology and Human Values* 19 (2). 189–204.

Midgley, M. (2002). *Beast and Man*. London: Routledge.

Midgley, M. (2004). *The Myths We Live By*. London: Routledge.

Midland Pig Producers Limited. (2011). *Foston Development*. Retrieved 4 July 2011, from Midland Pig Producers Limited: http://www.mppfoston.com/index.html.

Miller, K. S. and Knutson, J. F. (1997). Reports of Severe Physical Punishment and Exposure to Animal Cruelty by Inmates Convicted of Felonies and by University Students. *Child Abuse and Neglect 21 (1)*. 59–82.

Mills, C. W. (1970). *The Sociological Imagination*. Harmondsworth: Penguin.

Mullan, B. and Marvin, G. (1987). *Zoo Culture*. London: Weidenfeld and Nicolson.

Myers, O. E. (2003). No Longer the Lonely Species: A Post-Mead Perspective on Animals and Sociology. *International Journal of Sociology and Social Policy 23 (3)*. 46–68.

Natural England. (2011). *National Nature Reserves*. Retrieved 28 June 2011, from Natural England: http://www.naturalengland.org.uk/ourwork/conservation/designatedareas/nnr/default.aspx.

Nestler, E. J. (2001). Learning about Addiction from the Genome. *Nature 409 (15 February)*. 834–5.

Newton, T. (2007). *Nature and Sociology*. Oxon: Routledge.

Nibert, D. (2002). *Animal Rights/Human Rights: Entanglement of Oppression and Liberation*. Plymouth: Rowman and Littlefield.

Nibert, D. (2003). Humans and Other Animals: Sociology's Moral and Intellectual Challenge. *International Journal of Sociology and Social Policy 23 (3)*. 5–25.

Noel, D. L. (1972). *The Origins of American Slavery and Racism*. Columbus, OH: Merrill.

O'Connor, T. P. (1997). Working at Relationships: Another Look at Animal Domestication. *Antiquity 71*. 149–56.

O'Neill, J. (1972). *Sociology as a Skin Trade*. London: Heinemann Educational.

Owen, T. (2006). Genetic-Social Science and the Study of Human Biotechnology. *Current Sociology 54 (6)*. 897–917.

Parsons, T. (1940). An Analytical Approach to the Theory of Social Stratification. *American Journal of Sociology 45*. 841–62.

PAW Scotland. (16 March 2011). *PAW Scotland*. Retrieved 27 June 2011, from Illegal Trade CITES: http://www.scotland.gov.uk/Topics/Environment/Wildlife-Habitats/paw-scotland/types-of-crime/illegal-trade.

Peggs, K. (2009a). A Hostile World for Nonhuman Animals: Human Identification and the Oppression of Nonhuman Animals for Human Good. *Sociology 43 (1)*. 85–102.

Peggs, K. (2009b). The Social Constructionist Challenge to Primacy Identity and the Emancipation of Oppressed Groups: Human Primacy Identity Politics and the Human/'Animal' Dualism. *Sociological Research Online 14 (1)*. No pagination.

Peggs, K. (2010). Nonhuman Animal Experiments in the European Community: Human Values and Rational Choice. *Society and Animals 18 (1)*. 1–20.

Peggs, K. (2011a). Risk, Human Health and the Oppression of Nonhuman Animals: The Development of Transgenic Nonhuman Animals for Human Use. *Humanimalia: A Journal of Human/Animal Interface Studies. 2 (2)*. 49–69.

Peggs, K. (2011b). 'A call to arms'?: Exploring Sociology for Other Animals. London School of Economics. April 2011: British Sociological Association Annual Conference.

Philo, C. (1995). Animals, Geography, and the City: Notes on Inclusions and Exclusions. *Environment and Planning Development: Society and Space 13 (6)*. 655–81.

Philo, C. and Wilbert, C. (2000). Animal Spaces, Beasty Places: An Introduction. In C. Philo and C. Wilbert (Eds), *Animal Spaces, Beasty Places: New Geographies of Human Animal Relations* (pp. 1–36). London: Routledge.

Pink, S. (1997). *Women and Bullfighting: Gender, Sex and the Consumption of Tradition.* Oxford: Berg.

Planhol, X. (2004). *Le Paysage Animal. L'homme et la Grande Faune: Une Zoogéographie Historique.* Paris: Fayard.

Podberscek, A. L. (1994). Dog on a Tightrope: The Position of the Dog in British Society as Influenced by Press Reports on Dog Attacks (1988–1992). *Anthrozoos* 7. 232–41.

Podberscek, A. L. (2009). Good to Pet and Eat: The Keeping and Consuming of Dogs and Cats in South Korea. *Journal of Social Issues* 65 (3). 615–32.

Pro-Test. (2006). *About Us.* Retrieved July 2011, from Pro-Test: Standing up for Science: http://www.Pro-Test.org.uk/about.php.

Regan, T. (1983). *The Case for Animal Rights.* Berkeley: University of California Press.

Regan, T. (1985). The Case for Animal Rights. In P. Singer (Ed.), *In Defense of Animals* (pp. 13–26). New York: Basil Blackwell.

Regan, T. (1997). The Rights of Humans and Other Animals. *Ethics and Behaviour* 7. 103–11.

Regan, T. (2004). *Empty Cages: Facing the Challenge of Animal Rights.* Lanham, MD: Rowman and Littlefield.

Richardson, N. J., Macfie, H. J. H. and Shepherd, R. (1994). Consumer Attitudes to Meat Eating. *Meat Science* 36 (1). 57–65.

Ritvo, H. (1987). *The Animal Estate: The English and Other Creatures in the Victorian Age.* Cambridge, MA: Harvard University Press.

Ritzer, G. (1993). *The McDonaldization of Society.* Thousand Oaks, CA: Pine Forge Press.

Ritzer, G. and Malone, E. (2000). Globalization Theory: Lessons from the Exportation of McDonaldization and the New Means of Consumption. *American Studies* 41 (2/3). 97–118.

Rowlands, M. (2002). *Animals Like Us.* London: Verso.

RSPCA Australia. (2008). *How Are Beef Cattle Reared?* 25 November 2008. Retrieved 27 June 2011, from RSPCA Australia Knowledgebase: http://kb.rspca.org.au/How-are-beef-cattle-reared_88.html.

Russell, N. (2002). The Wild Side of Animal Domestication. *Society & Animals* 10 (3). 285–302.

Russett, C. E. (1989). *Sexual Science: The Victorian Construction of Womanhood.* Cambridge, MA: Harvard University Press.

Ryder, R. D. (1983 [1975]). *Victims of Science: The Use of Animals in Research. Revised edn.* London: National Anti-Vivisection Society.

Salt, H. H. (1980 (orig 1892)). *Animal Rights: Consideration in Relation to Social Progress.* London: Centaur.

Sanders, C. (1999). *Understanding Dogs.* Philadelphia: Temple University Press.

Sanger, J. T. (1996). *The Compleat Observer?* London: The Falmer Press.

Saul, J. M. (2003). *Feminism: Issues and Arguments.* Oxford: Oxford University Press.

Schroeder, J. E. (1998). Consuming Representation: A Visual Approach to Consumer Research. In B. B. Stern (Ed.), *Representing Consumers: Voices, Views and Visions.* (pp. 193–230). London: Routledge.

Shapiro, K. (1989). The Death of an Animal: Ontological Vulnerablility. *Between the Species* 5. 183–93.
Shapiro, K. (1995). The Caring Sleuth: A Qualitative Analysis of Animal Rights Activists. *Alternative Methods in Toxicology* 11. 669–74.
Shils, E. (1962). *Center and Periphery: Essays in Macrosociology*. Chicago: Chicago University Press.
Singer, P. (1976). *Animal Liberation. Towards an End to Man's Inhumanity to Animals*. London: Jonathan Cape.
Singer, P. (1990). *Animal Liberation. 2nd edn*. New York: New York Review of Books.
Singer, P. (1993). *Practical Ethics*. Cambridge: Cambridge University Press.
Singer, P. (2002). *Unsanctifying Human Life* (Ed. H. Kuhse). Oxford and Malden, MA: Blackwell.
Smart, B. (1999). *Resisiting McDonalidization*. London: SAGE.
Smart, B. (2005). *The Sport Star: Modern Sport and the Cultural Economy of Sporting Celebrity*. London: SAGE.
Smart, B. (2010). *Consumer Society: Critical Issues & Environmental Consequences*. London: SAGE.
Smith, D. E. (1987a). *The Everyday World As Problematic*. Boston, MA: Northeastern University Press.
Smith, D. E. (1987b). Women's Perspective as a Radical Critique of Sociology. In S. Harding (Ed.), *Feminism and Methodology* (pp. 84–96). Indiana: Indiana University Press.
Solot, D. (1997). Untangling the Animal Abuse Web. *Society and Animals* 5. 257–65.
Stewart, S. (2010). *Culture and the Middle Classes*. Farnham: Ashgate.
Stibbe, A. (2001). Language, Power and the Social Construction of Animals. *Society and Animals* 9 (2). 145–61.
Storey, J. (1993). *An Introductory Guide to Cultural Theory and Popular Culture*. London: Harvester Wheatsheaf.
Strydom, P. (2002). *Risk, Environment and Society: Ongoing Debates, Cultures, Issues and Future Prospects*. Buckingham: Open University Press.
Sutcliffe, F. E. (1968). Introduction. In R. Descartes (Ed.), *Discourse on Method and Other Writings* (pp. 7–23). Middlesex: Penguin.
Sykes, G. M. and Matza, D. (1957). Techniques of Neutralization: A Theory of Delinquency. *American Sociological Review* 22 (6). 664–70.
Szerszynski, B., Lash, S. and Wynne, B. (1996). Introduction: Ecology, Realism and the Social Sciences. In S. Lash, B. Szerszynski and B. Wynne (Eds), *Risk, Environment and Modernity: Towards a New Ecology* (pp. 1–26). London: SAGE.
Taylor, K. Gordon, N., Langley, G. and Higgins, W. (2008). Estimates of Worldwide Laboratory Animal Use in 2005. *Alternatives to Laboratory Animals* 36. 327–42.
Taylor, N. and Signal, T. (2008). Throwing the Baby out with the Bathwater: Towards a Sociology of the Human-Animal Abuse 'Link'? *Sociological Research Online* 13 (1). No pagination.
Tester, K. (1991). *Animals in Society: The Humanity of Animal Rights*. London: Routledge.
Tester, K. (1999). The Moral Malaise of McDonaldization: The Values of Vegetarianism. In B. Smart (Ed.), *Resisiting McDonaldization*. (pp. 207–21). London: Sage.

The Vegetarian Society. (2010). *Meat Consumption UK*. November 2010. Retrieved 30 March 2011, from Fact Sheets: http://www.vegsoc.org/page.aspx?pid=756.
Thomas, K. (1983). *Man and the Natural World: Changing Attitudes in England 1500–1800*. London: Allen Lane.
Thomas, W. I. and Thomas, D. S. (1928). *The Child in America*. New York: Knopf.
Tingle, D., Barnard, G. W., Robbins, L., Newman, G. and Hutchinson, D. (1986). Childhood and Adolescent Characteristics of Pedophiles and Rapists. *International Journal of Law and Psychiatry 9*. 103–16.
Traditional Bowhunter. (2011). *Traditional Bowhunter Online*. Retrieved 10 July 2011, from Traditional Bowhunter: http://www.tradbow.com/
Tuan, Y. (1984). *Dominance and Affection: The Making of Pets*. New Haven: Yale University Press.
Tulloch, J. and Lupton, D. (2003). *Risk and Everyday Life*. London: SAGE.
Turner, B. S. (2007). Culture, Technologies and Bodies: The Technological Utopia of Living Forever. *The Sociological Review 55 (1)*. 19–36.
Twine, R. (2010). *Animals as Biotechnology – Ethics, Sustainability and Critical Animal Studies*. London: Earthscan.
Twining, H. and Arluke, A. (2000). Managing the Stigma of Outlaw Breeds: The Case of Pit Bull Owners. *Society and Animals 8*. 1–27.
Tyler, T. (2006). An Animal Manifesto. Gender, Identity, and Vegan-Feminism in the Twenty-First Century. Carol J. Adams Interviewed by Tom Tyler. *Parallax 38 (Jan–Mar)*. 120–8.
Understanding Animal Research. (2011). *About Research*. Retrieved 11 July 2011, from Understanding Animal Research: http://www.understandinganimalresearch.org.uk/homepage.
United States Department of Agriculture. (2011). *Livestock and Poultry: World Markets and Trade*. United States Department of Agriculture.
Urry, J. (1990). *The Tourist Gaze: Leisure and Travel in Contemporary Societies*. London: SAGE.
Urry, J. (2011). *Climate Change and Society*. Cambridge: Polity.
Van Dijk, T. A. (1997). *Discourse as Social Interaction*. London: SAGE.
Veblen, T. (1994). *Theory of the Leisure Class: An Economic Study in the Evolution of Institutions*. London: Penguin.
Veevers, J. E. (1985). The Social Meaning of Pets: Alternative Roles for Companion Animals. *Marriage and Family Review 8 (34)*. 11–30.
Vialles, N. (1994). *Animal to Edible*. Cambridge: Cambridge University Press.
Wallis, L. (2011). *Vegan Anthems*. Spring 2011. Retrieved 23 July 2011, from The Vegan: http://www.vegansociety.com/feature-articles/vegan%20anthems.pdf.
Weber, M. (1978 (orig 1925)). *Economy and Society: An Outline of Interpretive Sociology* (Ed. G. Roth and G. Wittich). Berkeley: University of Calornia Press.
Wildlife Trust of India. (2007). *Wildlife Corridor Gives Endangered Elephants in India Passage between Reserves*. 21st December. Retrieved 21 June 2011, from ScienceDaily: http://www.sciencedaily.com /releases/2007/12/071220212827.htm.
Wilkie, R. (2005). Sentient Commodities and Productive Paradoxes: The Ambiguous Nature of Human–Livestock Relations in Northeast Scotland. *Journal of Rural Studies 21 (2)*. 213–30.

Wilkie, R. M. (2010). *Livestock/Deadstock: Food Animals, Ambiguous Relations, and Productive Contexts: Working with Farm Animals from Birth to Slaughter.* Philadelphia: Temple University Press.

Williams, B. (2006). *Philosophy as a Humanistic Discipline.* Princetown: Princetown University Press.

Williams, R. (1981). *Culture.* London: Fontana.

Williams, R. (1988). *Keywords.* London: Fontana.

Willsher, K. (2011). What is French for a Vegan? 30th March *The Guardian.*

Wilson, E. O. (1978). *On Human Nature.* Cambridge, MA: Harvard.

Wolch, J. (1998). Zoopolis. In J. Wolch and J. Emel (Eds), *Animals Geographies: Place, Politics, and Identity in the Nature-Culture Borderlands* (pp. 119–38). London: Verso.

Wolch, J. and Emel, J. (1998). *Animal Geographies: Place, Politics and Identity in the Nature-Culture Borderlands.* London: Verso.

Wolfe, C. (2010). *What is Posthumanism.* Minneapolis: University of Minnesota Press.

World Wildlife Fund. (2009). *Ten to Watch in 2010.* Retrieved 27 June 2011, from World Wildlife Fund: http://www.wwf.org.uk/wwf_articles.cfm?unewsid=3618.

Wright, J. A. and Hensley, C. (2003). From Animal Cruelty to Serial Murder: Applying the Graduation Hypothesis. *International Journal of Offender Therapy and Comparative Criminology 47 (1).* 71–88.

Yarwood, R. and Evans, N. (1998). New Places for 'Old Spots': The Changing Geographies of Domestic Livestock Animals. *Society and Animals 6 (2).* 137–65.

YouGov. (May 2006). *Daily Telegraph Survey Results.* Retrieved 25 June 2008, from YouGov: http.//:www.YouGov.com.

Author Index

Adams, Carol, 44, 93, 105, 118, 121, 150, 153, 164
Agamben, Michael, 42
Alexander, Victoria, D., 110, 112, 153
Alger, Janet, 5, 6, 10, 22, 31, 108, 153
Alger, Steven, 5, 6, 10, 22, 31, 108, 153
Allen Li, Jui-Chung, 157
Anderson, Kay, 78, 153
Anderson, Patricia K., 78, 153
Arkow, Phil, 63
Arluke, Arnold, 6, 29, 31, 45, 47, 74, 76, 77, 78, 80, 81, 82, 83, 84, 87, 88, 149, 153, 154, 164
Armstrong, Susan Jean, 18, 153, 156
Ascione, Frank R., 62, 153, 154, 157

Baker, Steve, 13, 110, 112, 113, 115, 117, 122, 123, 124, 126, 139, 140, 141, 153
Baralt, Lori, 120, 121, 159
Barnard, George W., 62, 164
Barthes, Roland, 113, 153
Bauman, Zygmunt, 90, 103, 129, 132, 144, 153
Beck-Gernsheim, Elizabeth, 132, 154
Beck, Ulrich, 35, 67, 71, 127, 129, 132, 139, 154
Beirne, Piers, 12, 49, 50, 51, 56, 57, 58, 64, 131, 154, 155
Bekoff, Marc, 108, 154
Bentham, Jeremy, 56, 154
Benton, Ted, 37, 56, 67, 131, 133, 134, 136, 154
Berger, John, 78, 111, 115, 124, 125, 154
Berger, Peter, 3, 4, 28, 154
Bergman, Charles, 85, 154
Biersted, Robert, 3, 154
Birke, Lynda, 80, 82, 131, 154, 160
Blaney, Don, 95, 154
Blumer, Herbert, 6, 7, 154
Bokonyi, Sandor, 75, 154

Botzler, Richard, 18, 153, 156
Bourdieu, Pierre, 33, 154
Broglio, Ron, 79, 160
Brown, Nik, 130, 154
Bryant, Clifton D., 2, 10–11, 14, 16, 58, 107, 145, 146, 147, 148, 149, 151, 154
Buller, Henry, 70, 154
Burawoy, Michael, 148, 149, 155
Burridge, Joseph, 109, 155
Busfield, Joan, 129, 155
Butler, Judith, 43, 155

Calarco, Matthew, 42, 155, 156
Calhoun, Craig, J., 34, 155
Carruthers, Peter, 138
Catton, William R., 67, 156
Cavalieri, Paola, 138, 155
Cazaux, Geertrui, 59, 155
Cohen, Robin, 3, 155
Cohen, Stan, 60, 155
Cohn, P. Michael, 131, 155
Cole, Matthew, 100, 105, 155
Collins, Patricia Hill, 44
Collins, Randall, 140, 155
Comte, Auguste, 3, 5, 6, 148, 155
Connell, Alex, 105, 155
Craig, Ian W., 27, 156
Crawford, Neta C., 144, 156
Cudworth, Erica, 150, 156

Darnton, Robert, 113, 156
Darwin, Charles, 22, 117, 156, 158
Davis, Christopher G., 100, 156
Davis, Kingsley, 100, 156
de Botton, Alain, 107, 156
Delgado, Christopher L., 93, 98, 156
Derrida, Jacques, 40, 42, 55, 113, 125, 142, 144, 155, 156
Descartes, Rene, 12, 17–20, 24, 25, 26, 31, 53, 56, 117, 156, 163
Digard, Jean-Pierre, 75, 156

Donovan, Josephine, 105, 156
Douglas, Mary, 82, 156
Douglass, Carrie, 120, 156
Ducos, Pierre, 75, 156
Duneier, Mitchell, 23, 158
Dunlap, Riley, 67, 156
Dupre, John, 17, 19, 20, 21, 22, 117, 156
Durkheim, Emile, 20, 24, 25, 26, 47, 52, 55, 69, 113, 114, 140, 156, 157

Emel, Jody, 70, 71, 86, 153, 157, 165
Engels, Friedrich, 34, 35, 66, 67, 160
Evans, Ewdard Payson, 61
Evans, Nick, 72, 165
Evans, Rhonda D., 61, 157

Faver, Catherine, 63, 157
Featherstone, Mike, 91, 157
Ferguson, Christopher J., 51, 157
Finelli, Mary, 86, 160
Fishman, Joshua A., 42, 157
Fitzgerald, Amy, 18, 54, 120, 121, 156, 159
Flynn, Clifton, P., 12, 57, 60, 61, 62, 63, 64, 157
Forsyth, Craig J., 61, 157
Foucault, Michel, 43, 71, 118, 157
Francione, Gary L., 18, 19, 105, 157
Francis, Fay, 97, 98, 157
Franklin, Adrian, 68, 76, 79, 86, 87, 92, 93, 94, 95, 96, 98, 99, 100, 101, 120, 125, 128, 135, 138, 141, 157
Freese, Jeremy, 20, 157
Frost, David, 97, 98, 157
Fuller, Steve, 5, 16, 20, 28, 30, 42, 142, 157

Garner, Robert, 53, 54, 94, 97, 132, 133, 134, 135, 136, 137, 138, 141, 143, 157
Gauthier, Deann K., 61, 157
Geertz, Clifford, 108, 157
Gerteis, Joseph, 155
Giddens, Anthony, 5, 23, 39, 41, 129, 140, 158
Goodall, Peter, 108, 158
Gordon, N., 128, 164

Goring, Charles, 50, 158
Gouldner, Alvin, 148, 149, 158
Gould, Stephen J., 50, 116, 158
Grant, Catharine, 128, 158
Gray, John, 131, 158
Guo, Guang, 20, 22, 158

Hafferty, Frederic, 6, 153
Hall, Stuart, 40, 41, 43, 110, 154, 158
Hammersley, Martyn, 148, 150, 151, 158
Haraway, Donna, 29, 70, 158
Harding, Sandra, 150, 158, 163
Henry, Bill, 131, 158
Hensley, Christopher, 61, 165
Hickrod, Lucy, 76, 158
Higgins, W., 128, 163
Hinchliffe, Steve, 73, 158
Holloway, Lewis, 97, 158
Hooks, bell, 44, 52
Humphries, Beth, 5, 149, 158
Hutchinson, David, 62, 164

Ingold, Timothy, 86, 122, 158
Irvine, Leslie, 8, 9, 30, 31, 108, 158, 159

Jasanoff, Sheila, 27, 158
Jenkins, Richard, 41, 43, 159
Jerolmack, Colin, 69, 81, 82, 159

Kalof, Linda, 18, 54, 120, 121, 122, 156, 159
Katz, Donald R., 119
Kean, Hilda, 85, 159
Kemp, Martin, 116, 117, 159
Kennedy, Paul, 3, 155
Klein, Naomi, 91, 159
Knight, Andrew, 119, 130, 159
Knutson, John F., 62, 161
Kruse, Corwin R., 2, 6, 10, 159

Laclau, Ernesto, 40, 142, 159
Langley, G., 128, 164
Lash, Scott, 68, 70, 164
Latour, Bruno, 69, 71, 159
Lawrence, Elizabeth A., 72, 159
Leavis, Queenie Dorothy, 109, 159
Lefebvre, Henri, 68, 159

Levi-Strauss, Claude, 101, 114, 115, 159
Lilienfeld, Scott O., 79, 160
Lin, Biing-Hwan, 100, 156
Linzey, Andrew, 58, 118, 136, 159
Lloyd, Moya, 44, 142, 159
Lombroso, Cesare, 50, 116
Lovegrove, 83, 159
Low, Martina, 69, 159
Luckmann, Thomas, 28, 154
Lupton, Deborah, 99, 129, 159, 164
Lynch, Michael, 80, 159
Lyons, Dan, 131, 159

Macfie, H. J. H., 101, 162
Macionis, John, 1, 3, 160
MacNair, Rachel M., 101, 160
Maher, Jennifer, 60, 160
Malamud, Randy, 78, 79, 124, 125, 160
Malone, Elizabeth, 102, 162
Marino, Lori, 79, 160
Marvin, Garry, 87, 120, 124, 125, 160, 161
Marx, Karl, 20, 34, 35, 66, 67, 160
Marx, Leo, 67, 160
Mason, Jim, 86, 160
Massey, Doreen, 68, 160
Matza, David, 61, 163
Maurer, Donna, 139, 160
McDonald, Barbara, 101, 160
McKibben, Bill, 68, 160
Mead, George, Herbert, 2, 6, 7–10, 12, 16, 18, 20, 24, 30, 40, 42, 108, 148, 153, 158, 160
Merchant, Carolyn, 68, 160
Michael, Mike, 42, 80, 82, 130, 131, 148, 154, 160
Midgley, Marry, 42, 47, 116, 117, 160
Miller, Karla S., 62, 161
Mills, Charles W., 3, 4, 39, 74, 161
Moody, James, 155
Moore, Wilbert, 34, 156
Morgan, Karen, 100, 105, 155
Morris, Carol, 70, 154
Mullan, Bob, 124, 125, 161
Myers, Olin, 2, 161

Nestler, Eric J., 27, 166
Newman, Gustave, 62, 164
Newton, Tim, 27, 69, 161
Nibert, David, 12, 14, 33, 36, 37, 38, 39, 42, 151, 161
Nobis, Nathan, 79, 160
Noel, Donald L., 37, 161

O'Connor, Terry, 75, 161
O'Neill, John, 2, 161
Owen, Tim, 28, 161

Parker, James, V., 131, 155
Parsons, Talcott, 34, 161
Payson Evans, 52
Peggs, Kay, 14, 41, 42, 43, 128, 130, 131, 137, 143, 144, 147, 149, 150, 161
Pfaff, Steven, 155
Philo, Chris, 65, 71, 72, 80, 82, 84, 88, 161, 162
Pierpoint, Harriet, 60, 160
Pink, Sarah, 120, 162
Planhol, Xavier de, 70, 162
Plummer, Ken, 1, 3, 160
Podberscek, Anthony, 59, 77, 162
Pulcino, Roarke, 131, 158

Rawls, John, 137
Regan, Tom, 19, 54, 93, 97, 133, 138, 162
Richardson, N. J., 101, 162
Ritvo, Harriet, 51, 55, 162
Ritzer, George, 13, 90, 91, 102, 103, 162
Robbins, Lynn, 62, 164
Rousseau, Jean-Jacques, 115
Rowlands, Mark, 17, 19, 55, 56, 87, 137, 138, 141, 143, 162
Russell, Nerissa, 74, 162
Russett, Cynthia, 23, 162
Ryder, Richard, 36, 162

Salt, Henry, 48, 135, 162
Sanders, Clinton, 6, 29, 30, 31, 45, 47, 74, 76, 77, 78, 80, 81, 82, 83, 84, 87, 108, 153, 162
Sanger, Jack, 1, 151, 162
Saul, Jennifer Mather, 23, 162

Schmidt, Kathryn, 155
Schmitt, Raymond, 76, 158
Schroeder, Jonathan E., 119, 162
Shapiro, Kenneth, 31, 93, 163
Shepherd, R., 101, 162
Shils, Edward, 34, 163
Signal, Tania, 62, 163
Singer, Peter, 36, 54, 56, 57, 92, 103, 135, 136, 137, 138, 160, 162, 163
Smart, Barry, 91, 102, 119, 163
Smith, Carolyn, 97, 98, 157
Smith, Dorothy, 149, 150, 163
Solot, Dorian, 63, 163
South, Nigel, 64, 77, 154, 155, 162
Stewart, Simon, 108, 110, 163
Stibbe, Arran, 92, 94, 95, 163
Storey, John, 110, 163
Strand, Elizabeth, 63, 157
Strydom, 129, 163
Sutcliffe, 18, 25, 163
Sykes, Gresham M., 61, 163
Szerszynski, Bronislaw, 68, 70, 163

Taylor, Katy, 128, 163
Taylor, Nik, 62, 163
Tester, Keith, 29, 99, 103, 111, 133, 134, 135, 138, 141, 163
Thomas, Dorothy Swaine, 50, 164
Thomas, Keith, 71, 76, 164
Thomas, William Isaac, 50, 164
Tingle, David, 62, 164
Tuan, Yi-Fi, 72, 77, 84, 164
Tulloch, John, 129, 164
Turner, Brian, 27, 130, 164

Twine, Richard, 95, 164
Twining, Hilary, 6, 164
Tyler, Tom, 105, 164

Urry, John, 67, 118, 164

Van Dijk, Tuen, A., 92, 164
Veblen, Thorstein, 91, 164
Veevers, Jean E., 63, 164
Vialles, Noélie, 100, 164
Virk, Indermohan, 155

Wade, Lisa D., 157
Wallis, Louise, 110, 164
Weber, Claudia V., 63, 153
Weber, Max, 2, 4, 6, 20, 30, 35, 39, 108, 164
Welchman, Jane, 131
Wilbert, Chris, 34, 72, 80, 82, 84, 86, 88, 157, 162
Wilkie, Rhoda, M., 94, 96, 164, 165
Williams, Bernard, 130, 165
Williams, Raymond, 108, 165
Willsher, Kim, 100, 165
Wilson, Edward O., 22, 23, 165
Wolch, Jennifer, 66, 70, 71, 72, 73, 81, 83, 86, 153, 157, 165
Wolfe, Cary, 150, 165
Wood, David S., 63, 153
Wright, Jeremy, 61, 165
Wynne, Brian, 68, 70, 163

Yarwood, Richard, 72, 165

Subject Index

abbattoirs, *see* slaughter houses
abuse, 12, 13, 46, 47, 49, 53, 55, 56, 57, 58, 59, 60, 61, 62, 63, 64, 77, 131, 141, 143, 146, 150
action, 6–7
activism, 100
   animal rights, 14, 140, 141, 146, 149
   grass roots, 13, 127, 141
age, 34, 41, 95
agency, 72, 111, 138
agricultural industries, 92
agriculture, 56, 71, 93, 99, 131, 134
American Pit Bull, 59
American Sociological Association, 6
analytic animal, 80
animal abuse and cruelty, 12, 13, 45–64, 77, 94, 101, 102, 112, 120, 131, 141, 143, 146, 150
animal advocacy
   contractarianism, 53, 55, 137–8
   rights, 14, 53, 132–5, 137, 138–41, 143–4
   utilitarianism, 53, 54, 57, 67, 135, 136
animal experiments, 14, 19, 38, 43, 63, 78, 80–1, 127–32, 133, 134, 138, 139, 140, 141–4, 147
animal health, 95, 97
Animal Health and Welfare (Scotland) Act 2006, 57
animalization, 12, 47
animals
   as automata, 17–19, 31, 56, 64, 77, 117
   as 'bad', 74, 81–3, 87–8
   as criminals, 51–2
   as 'good', 74, 76–81, 84–7
   as property, 11, 38, 52, 58–9, 140
   as symbols, 38
   as witnesses, 51
animal selves, 30–1
animal studies, 149

Animal Welfare Act (UK) 2006, 56, 57
Animal Welfare Act (US) 1966, 57
anthropocentrism, 64, 67, 79, 125, 146
anthropologists, 113
anthropomorhism, 9, 11, 21, 64, 67, 77, 79, 121, 125, 146
antibiotics, 97
anti-speciesism, 105
antithesis, 26, 78, 82
aquariums, 78, 79, 88
Armani, 91, 118
arrogant eye, 118, 119
arts, the, 38, 112, 113, 124
Australia, 96, 128, 162

Bacon, 92, 95
Badgers, 70, 73
battery eggs, 95
beef, 93, 96, 97, 100, 102, 109, 126, 163
beef belt, 96
binary classifications, 16–17, 40, 69, 142
biological determinism/essentialism, 16, 22, 23, 31
biology, definition of, 20
biomedical, 38, 54, 128, 130, 131
birds, 9, 31, 56, 62, 67, 108
blackbirds, 65
Bosch, Hieronymus, 116, 126
boundary work, 45, 51, 73
bourgeoisie, 34, 66
bovine spongiform encephalopathy (BSE), 99
boxing, 61
boycott, 96, 104
branding, 59, 91
Britain, 51, 60, 65, 85, 87, 88, 90, 97, 102, 116, 120, 158, 159
British Broadcasting Corporation (BBC), 107, 140, 154, 157

170

British Government, 60
British Hen Welfare Trust, 97
British Sociological Association, 6, 162
bulldogs, 29
bulls, 83, 93, 120, 121, 124
burden of proof, 30
butchery, 86, 101
butterflies, 85

calves, 86, 93, 94
Cambodia, 77
cancer, 98, 99, 129
capitalism, 84, 111
caring role, 22
Carmen, 107
Cartesian, 19, 64, 138
cats, 30, 38, 51, 65, 67, 72, 77, 80, 108
chickens, 8, 56, 86, 87, 95, 99, 102
childhood, 61, 62
choice, 4, 14, 90, 100, 105, 137
Christ Taken Captive (painting), 116
civilized, 24, 26, 27, 50, 66, 120
class, 12, 33, 34–5, 40, 41, 44, 59, 63, 109, 110, 137
classism, 36, 37, 40
climate change, 84
cocks, 51, 146
  fighting, 146
cognitive, 29
collective conscience, 26
commodification, 106, 130
commodities, 58, 64, 91, 105, 130
communication, 8, 19, 31, 108
  'conversation of gestures', 8
  'significant symbols', 8–9
  see also language use
companion animals, 38, 60, 62, 70, 72, 76, 111
  see also pets
compassion, 101, 131, 138, 141
Compassion in World Farming (CWF), 94, 138, 155
conflict, 9, 26, 34, 35, 58, 123, 143
consciousness, 18, 26, 29, 30
conservation, 79, 84, 87, 89, 122, 124, 162
conspicuous consumption, 91
constructionism, *see* social constructionism

consumerism, 106
consumer s, 13, 86, 90–106, 110, 121, 123, 134
consumer societies, 90
continuum of recognition, 53, 132
control, 37, 39, 49, 57, 69, 73, 75, 84, 87, 102, 103, 104, 122, 129, 130, 137
Convention on International Trade in Endangered Species (CITES), 91
cooking, 101
cosmetics, 38
Countryside Alliance UK, The, 109, 139, 155
cows, 48, 56, 65, 93, 95, 96, 98, 99, 100, 102, 105
Creutzfeldt-Jakob disease(CJD), 99, 129
crime
  and morality, 47–9, 52
  as social construction, 49
criminology, 49, 50, 57, 58, 59, 64
  definition, 49
  as a discourse, 57
cruelty, 48, 56, 57, 58, 61, 104, 131
cuisine, 97, 124
cultural representations, 117, 126
cultural studies, 109, 110
culture, 20, 41, 107–8, 113, 153, 154, 156–61, 163, 165
  general usage, 108, 109, 111, 126
  restrictive usage, 108, 109

dairy, 92, 101
Dangerous Dogs Act (UK) 1991, 60, 83
death of nature, 68
decoding, 110, 111, 120
deer, 56, 65, 73, 120, 121
deforestation, 84
demons, 81, 83
desensitization thesis, 62
deviance, 47–9, 52, 157
  as social construction, 49
diet, 93, 96, 97, 100, 102, 105
dirt, 82
discourse, 42, 43, 49, 57, 92–3, 115, 120
  role in social construction, 43
Disney Corporation, 123

disnification, 13, 123, 124, 126
distancing concepts, 10, 23
distinction, 7, 19, 21, 24, 26, 29, 32, 36, 42, 43, 65, 69, 71, 73, 85, 115, 117, 148
divorce, 63
dogmen, 61
Dogo Argentino, 83
dogs, 8, 18, 30, 38, 51, 59, 60, 61, 62, 65, 67, 72, 77, 80, 83, 109, 129, 139, 140
 attacks, 59, 60
 dolphins, 65, 146, 147
 fighting, 38, 61
 racing, 38, 146
 status, 47
 weapons, 47
domesticatation, 13, 42, 65, 70, 74–8, 80, 82, 86–7
domesticatory action, 75
domestic violence, 62–3
donkeys, 1, 152
DreamWorks, 124
drugs, 27, 97, 128
dualism, 26, 70, 71
 body/soul, 25, 26
 human/animal, 28, 69–72, 161
 nature/culture, 66, 69–72, 71, 74
Du Pont, 80

eagles, 73
early modern, 67, 76
ear tagging, 59
ecology, 67, 71, 87, 141
elephants, 23, 74
elites, 39
emotions, 38, 58, 63, 96, 98, 103, 153, 154, 155
 emotional attachment, 63, 94
 emotional detachment, 94
emus, 98
encoding, 110
endangered animals, 42, 79, 84–6, 91, 125, 154, 164
England, 48, 73, 76, 85, 161, 164
essentialism, 21, 23, 32
ethnicity, 34, 41, 137
evolution, 9, 22, 48, 50, 52, 117

exclusion, 20, 35, 40, 41, 42, 45, 49, 57, 72, 142
exploitation, 32, 35, 37–9, 44, 57, 66, 73, 75, 84, 88, 92, 102, 105, 133, 134, 143, 144, 150, 151

factory farming, 104
family, 25, 62, 63, 76, 77, 88
farm animals, 70, 75
farming, 13, 38, 47, 63, 64–5, 71–3, 86, 88–9, 93–9, 100, 103–6, 118, 133, 155
feminism, 23, 105, 118, 142, 150
Fila Braziliero, 83
fish, 56, 92, 93, 98, 146
folk devils, 60
Fordism, 13, 94, 96, 99, 102, 104
foxes, 87, 88
foxhounds, 88
frame breaks, 76
France, 17, 24, 51, 100, 109, 124, 128, 156, 162
freedom fighters, 43
free range, 95
French, 17, 24, 109, 124, 156, 173
fur, 90, 91, 105, 118, 119
 farming, 118
 industry, 118

gannets, 65
Gaultier, 91, 118
gaze, 58, 99, 100, 118, 119, 125
gelatine, 90
gender, 2, 12, 17, 20, 21, 23, 33, 34, 36, 41, 43–4, 118, 120, 137, 155, 159, 162, 164
gendered, 23, 44, 120, 121
gene, 27, 28
generalized other, 9
genetics, 12, 21, 22, 23, 27, 28, 29, 42, 50, 51, 81, 95, 128, 130
 as explanation for crime, 51
genome, 27
Germanic jurisprudence, 51
German shepherd dog, 59
goats, 95
gorillas, 42, 54, 85
graduation hypothesis, 62
Great Ape Project, The, 54

great apes, 42, 54
guard dogs, 78
gulls, 65

hamsters, 93
heart disease, 98, 99, 129
hegemony, 110
hens, 95, 97
heritage, 73, 85, 85
historical, 66, 79, 88, 125
horse racing, 38, 120
horses, 13, 18, 38, 59, 71, 107, 120, 146
House of Lords (UK), 48
human health, 96, 98–102, 104, 106, 129–32
humanitarian, 52
humanitarianism, 52
human as non animal, 17
human rights, 52, 133
hunting, 14, 38, 56, 87–8, 107, 109, 120, 121, 135, 139, 146
Hunting Act (UK) 2004, 88

identification, 40, 41, 141
identity, 14, 27, 40–4, 45–6, 91
 criminal, 60–3
 human, 111, 115–18, 126, 141–3
 national, 77, 85, 119–20
 primary, 43
identity markers, 91
identity politics, 141–3
 primacy, 142–3
ideological control, 37, 39–44
ideology, 36, 40, 159
images, 115–19, 122–6
imaginary, 26, 55
India, 57, 74, 98, 164
individualization, 35, 131
industrial farming, 38, 87, 94, 97, 106
 see also intensive farming
inequalities, 5, 12, 16, 21, 23, 32–9, 118, 129
 sociobiological explanations of, 23
innate, 28, 29, 151
instinct, 8, 9, 10, 19, 22, 29, 31
integration, 25
intelligence, 9, 22, 23, 117

intensive farming, 94, 95
 see also industrial farming
interhuman violence, 61–3
interlocking systems of domination, 44, 52
invertebrates, 20, 42, 128, 137
Ireland, 120
Irish Republican Army (IRA), 140
irrationalities, 103

Japan, 93
Japanese Tosa, 83
Jaws (film), 110

kangaroos, 98
Keep Britain Tidy 'Dirty Pig' campaign, 116
knowledge, 4, 28, 71, 137, 148, 149, 150
Korea Animal Rights Advocates (KARA), 139
Kung Fu Panda (film), 124

labelling theory, 49, 54
laboratories, 13, 47, 54, 56, 63–5, 72, 78–82, 89, 131, 133, 141
labour, 34, 37, 94, 95
ladder of worth, 74
language use, 7–10, 14–15, 18–19, 31, 41, 42–4, 45, 108, 115
 and power, 43
leather, 91, 92, 105, 135, 141
leisure, 13, 59, 64, 90, 91, 106, 107, 108, 145, 146
liberation, 79, 138, 140, 142, 144, 147
lifestyle, 35, 98, 100, 104, 105, 106, 141
Linda McCartney, 105, 155
literature, 11, 14, 34, 76, 109, 110, 115, 138
Little Red Riding Hood, 123
logos, 122, 124

male, 17, 23, 55, 62, 119, 120, 126, 149
malestream, 118
Martins Act (UK) 1822, 48
Marxism, 110
masculinity, 61, 120

materialism, 13, 90
  gross, 13, 90
Mayor of London, 82
McDonaldization, 13, 102–5, 162, 163
McDonald's, 13, 102, 103, 104, 106, 107, 160
McLibel trial, 102
McSpotlight, 102, 103, 160
meat, 4, 13, 14, 19, 86, 90, 91–106, 110, 135, 141
mediated, 122, 123, 140
medical, 118, 129, 130
medieval, 52
mentalité, 113, 115
mice, 70, 80, 130
Mickey Mouse, 123
Midland Pig Producers, 97, 160
milk, 13, 90, 92, 93, 94, 95, 98
minding animals, 108
minks, 91, 118
modernity, 81, 92
mods, 60, 155
monkeys, 18, 31, 51, 108, 147
moral
  agents, 54, 55
  authority, 114
  community, 54
  consideration, 49
  orthodoxy, 53, 132, 138, 143
  panics, 60
  patients, 55
  purpose of sociology, 148
  standing, 52, 53, 54, 132, 138, 143
  status, 53, 54, 132, 143
  worth, 53, 54
morality, 26, 48, 52, 53, 103–6, 140, 144, 148
murder, 47, 54, 56, 93

narratives, 120, 121
National Anti-Vivisection Society UK(NAVS), 127, 162
national boundaries, 73, 77, 83
natural
  environment, 67, 68, 69, 73
  resources, 68, 70
  rights, 134
  selection, 22
naturalisation, 113

naturalistic animal, 80
nature reserves, 73, 74
Nazism, 45
neutralization theory, 61
news media, 47, 59, 60, 100, 116, 123, 129, 140, 141
New York City Animal Rights, 139
norms, 48, 131
nurture, 32, 108, 111

object, 8, 9, 30, 118, 119
objectification, 119
objective, 25, 68, 148, 149, 150
observation (as methodological approach), 18
ocelots, 79, 125
OncoMouseTM, 80
ontological, 49, 51, 56, 70, 81, 82
oppressions, 33, 36, 37, 39, 44, 45, 46, 151
  links between oppressions, 44
orang-utans, 42, 54
organic, 44, 91, 97, 142
other, the, 39, 40, 44, 46, 51, 85, 88, 142

pain, 18, 19, 20, 49, 56, 57, 64, 95, 104, 136, 143
paintings, 13, 108, 110, 114, 115, 126
pandas, 85, 122
party
  as different to class, 35
pathologized, 51, 73, 99
patriarchy, 44, 119, 120, 121, 122, 142
People for the Ethical Treatment of Animals (PETA), 127, 139
pests, 87, 88
pets, 29, 38, 63, 66, 75, 76–8, 80, 81, 82, 83, 133, 134
  see also companion animals
pharmaceutical, 129, 151
philosophy, 58, 109
photographs, 90, 120
physiognomy, 116
pigeons, 65, 70, 72, 73, 82
pigs, 51, 56, 86, 93, 95, 97
pit bull terriers, 60, 61, 83
plants, 71, 73, 92, 128
Playboy bunnies, 146

poaching, 59, 84
polar bears, 79
police horses, 38
political parties, 35, 139
popular culture, 47, 110, 114
population, 22, 66, 67, 75, 81, 85, 92, 94, 96, 101
pornography, 121
positivism, 4, 50, 155
posthumanism, 150, 165
postmodern, 144
power, 35–46, 50, 51, 53, 59, 60, 67, 69, 71, 74, 77, 96, 98, 122, 134, 145
  relations, 12, 13, 33, 37, 39, 65, 77, 88, 92, 119, 142
  transformative capacity of, 39
prejudice, 36, 37, 54
private property, 38, 58
production, 35, 56, 69, 86, 91, 94, 95, 96, 97, 98, 99, 100, 102
profane, 113
profiling, 27
profit, 34, 38, 75, 94, 104, 130
progress, 2, 4, 59, 66, 70, 71, 78, 99, 117, 128, 145
progression thesis, 62
proletariat, 34
protection, 8, 38, 48, 49, 57, 60, 73, 86, 125, 128, 132, 134, 136
protection of interests, 134, 136
Pro-Test UK, 127, 139, 142, 143, 162
psychology, 24, 62
puffins, 65

race, 12, 33, 34, 36, 44, 59, 77, 107, 124
racism, 36, 37, 40, 44, 46
racist patriarchy, 44
rape, 47, 56, 62, 116
rare, 79, 84, 85, 125
Ratatouille (film), 124
rational, 9, 103, 105, 138
rationality, 103
  irrationality of, 102
rationalization, 102, 103, 106
rationalizing processes, 102
rats, 72, 73, 80, 82, 83, 85
realism, 68

referential communication, 31
reflexivity, 135, 140, 148
regulation, 25, 128
religious life, 25
Renaissance man, 109
representations of animals, 13, 69, 79, 107–8, 111–26
reptiles, 62
researchers, 27, 57, 80
resources, 37, 45, 93, 128, 131
responsibilities, 19, 51, 54, 134
rhinoceros, 84
rights, 14, 35, 43, 51–7, 132–48
risk, 12, 27, 47, 67, 85, 99, 127, 129
ritual, 31, 108, 113, 140, 160
robins, 65
rockers, 60, 156
rodents, 62, 80
rodeos, 146
rottweilers, 59
Royal Society for the Protection of Animals UK (RSPCA), 96, 139, 163
rural idyll, 65, 86, 100

sacred, 25, 113, 114
sacrifice, 26, 31, 81, 143
salmonella, 99
science, 4, 7, 17, 58, 68, 109, 116, 130
scientists, 49, 134
sculpture, 114
seals, 65
self, 7, 9, 19, 24, 27, 30–1, 40, 54, 104, 135, 144, 151
self-defence, 54
sentience, 11, 19, 20, 53, 64, 137
sexism, 36, 37, 40, 46
sexuality, 34, 122, 137
sexualization, 120
sharks, 110
sheep, 56, 59, 65, 93, 95, 98
shellfish, 93
shooting, 62, 109, 120
shopping, 90, 91
significant symbols, 8, 9
silk, 92
simianized images, 117
slaughter houses, 13, 86, 100
slavery, 121
Smiths, The, 110

sniffer dogs, 78
social
  divisions, 33–6
  facts, 25, 34
  inequalities, 12, 23, 33, 34, 35, 37
  interaction, 28
  movements, 39, 139–42
social construction
  animal, 28–9, 77, 78, 81
  animal rights, 141
  gender, 44
  human, 28–9
  nature, 70–3
  race, 40
social constructionism, 28–9, 42–3, 49, 51, 52, 68
socialization, 11, 101
social order, 35, 69
societies, 2
sociobiology, 22–4, 25, 27, 29
sociological analysis, 3, 11, 12, 16, 20, 33, 36, 58, 113
sociological imagination, 3, 4–5, 11, 20–1, 74, 144
sociologists, 2, 6, 9, 20, 24, 30, 37, 67, 150
sociology
  definitions of, 1, 3–4
  distinct from biology, 16, 20–1
  main focus, 2
  purpose of, 3–4, 14, 147–8
  role of meaning in, 6–7
  scientific study, 3, 4, 7
  textbooks, 1, 3, 4, 5–6, 10, 23
sociozoologic system, 74–88
soul, 18, 25, 26
South Korea, 77, 162
Soviet, 110
Spain, 107
SPEAK, 138, 139
speciesism, 36–7, 40, 44–6, 64, 104, 105, 136, 143, 151, 155
spectacle, 13, 78, 79, 111, 114, 146
speech, 2, 18, 19, 48, 57, 137
sport, 56, 59, 109, 114, 119, 120
squirrels, 65, 85, 86
standardized animals, 130
standpoint, 55, 149, 150, 151
Star, The, 140

state, 37, 49, 75, 82, 99, 100, 121
status food, 98
stereotypical, 120, 123
stigmata, 50
stratification, 12, 33, 34, 35, 37, 40, 41, 42, 45, 46
  as different to class, 34
  see also class
subject-of-a-life, 54, 133
subpolitics, 127, 140
suffering, 5, 26, 49, 51–64, 81, 87, 93, 94, 95, 97, 103, 105, 118, 131, 136, 143, 147, 150
sui generis, 24, 25, 26
superiority, 5, 10, 18, 22, 25, 34, 36, 40, 42, 45, 87, 117, 119, 124, 126
superobjects, 80
surrogate enemies, 63
survival, 22, 26, 85
swine flu, 129
symbiosis, 75
symbolic interaction/ism, 6–7, 28, 30–1
symbolism, 120
symbols, 12, 47, 82, 86, 111, 113, 114

Taiwan, 128
taming, 75
tattoos, 59
taxidermy, 146
technology, 67, 68, 70, 71, 75, 95, 103, 104
teddy bears, 124
television, 112, 114, 121, 124
terminology, 14, 54, 69, 76
terrorism, 43, 140–1
Thailand, 77
theme parks, 73
theology, 4, 58
theory of ethnic stratification, 37
theory of oppressions, 4, 58
  economic exploitation/competition, 37–9
  ideological control, 37, 39–44
  unequal power, 37, 39, 77
tigers, 29, 79, 84
Tony Blair, 117
totemism, 69, 114, 122, 160

tourists and tourism, 82, 93
  ecotourism, 88
trades unions, 139
tradition, 119, 120, 131
Traditional Bowhunter, 120, 121, 122, 164
traditional societies, 69
trapping, 118
tuna, 84, 146, 147

UK, 48, 53, 56, 57, 59, 60, 77, 80, 83, 85, 86, 93, 94, 97, 99, 101, 102, 105, 109, 110, 117, 128, 129, 130, 131, 139, 140, 142, 153, 155, 164
Understanding Animal Research, 130, 165
unequal power relations, 20, 77
universal, 8, 25, 26, 54, 91, 133, 143
urban environment, 13, 60, 65, 66, 67, 68, 72, 73, 74, 86
urbanization, 66, 67, 70, 71, 73, 74, 83
urban sociology, 66
urban theory, 66
USA, 61, 62, 77, 83, 94, 95, 96, 110, 117, 120, 128, 139, 156, 164
utility, 71, 78, 80, 88, 91, 135
utopian, 130

vaccines, 130
values, 7, 26, 48, 60, 77, 84, 103, 121–2, 131, 149, 150, 154, 160–1, 163
vandalism, 60
veal, 94, 96, 155
  veal crates, 94
veganism, 96, 100, 101–2, 105, 110, 138, 164
vegaphobia, 100
vegetarianism, 13, 91, 96, 99, 100, 101, 102, 103, 104, 105, 106
  ethical, 104
  life style, 105

Vegetarian Society UK, The, 93, 101, 105, 164
veil of ignorance, 55, 137
vermin, 72, 81, 82, 83, 87, 89
verminization, 82
vertebrate, 42, 128
Vietnam, 77
violence, 11, 12, 19, 47, 57–64, 100, 109, 112, 122
  institutionalized, 64
  web of, 63

Wales, 97, 158
wealth, 29, 35, 76, 90, 91, 93, 96, 98, 129
welfare, 58, 59, 94, 101, 103, 132, 143, 146, 147
West, The, 66, 77, 78, 79, 91, 96, 100, 104
whales, 146, 147
wildlife documentaries, 112, 124
Wildlife Trust of India, 74
Wild Mammals (Protection) Act UK 1996, 48
wine, 92
wolves, 8, 123
women, 20, 21, 23, 62, 63, 102, 110, 118, 119, 120, 121, 126, 134, 142, 146, 149, 150
women's rights, 134
wool, 86, 90, 92, 105
workforce, 94
World Wildlife Fund (WWF), 84, 122, 124, 165
'wronged subjects, animals as, 55–8

youth, 60

zoological connection, 11, 14, 16, 58–9, 107, 145–7, 148, 149, 151
zoological 'correction', 145, 147
zoos, 13, 38, 65, 67, 72, 78–80, 83, 107, 111, 124, 125, 133, 134, 146